T0135873

ANALYZING VARIOUS ASPECTS OF SCHEDULING INDEPENDENT JOBS ON IDENTICAL MACHINES

DISSERTATION

zur Erlangung des akademischen Grades
doctor rerum naturalium (Dr. rer. nat.)

vorgelegt dem Rat der Fakultät für Mathematik und Informatik
der Friedrich-Schiller-Universität Jena

von Diplom-Wirtschaftsmathematiker Rico Walter
geboren am 03.09.1981 in Hildburghausen

Jena, 16. Februar 2010

Bibliografische Information der Deutschen Nationalbibliothek

Die Deutsche Nationalbibliothek verzeichnet diese Publikation in der
Deutschen Nationalbibliografie; detaillierte bibliografische Daten sind
im Internet über http://dnb.d-nb.de abrufbar.

Zugl.: Diss., Univ. Jena, 2010

ISBN 978-3-8325-2550-7

Logos Verlag Berlin GmbH
Comeniushof, Gubener Str. 47,
10243 Berlin
Tel.: +49 (0)30 42 85 10 90
Fax: +49 (0)30 42 85 10 92
INTERNET: http://www.logos-verlag.de

Contents

Deutsche Zusammenfassung

Auf dem Gebiet der Angewandten Mathematik und des Operations Research haben sich Scheduling-Probleme in den letzten Jahrzehnten zu einem der wichtigsten Forschungsschwerpunkte entwickelt. Neben grundlegenden theoretischen Ergebnissen gibt es zahlreiche Veröffentlichungen zu experimentellen Resultaten - gewonnen durch intensive Computersimulationen - insbesondere im Zusammenhang mit der Frage nach der Güte von heuristischen Lösungsverfahren.

Scheduling-Probleme sind aber nicht nur von akademischem Interesse, sondern auch von großer praktischer Relevanz, da sie in vielen Bereichen auftreten. Typische Anwendungsgebiete finden sich in der Industrie (z. B. Maschinenbelegungs-Planung), in der Informatik (z. B. Zuordnung von Jobs zu Prozessoren) sowie beispielsweise bei der Erstellung von Stundenplänen an Schulen und Universitäten.

Die vorliegende Arbeit zum Thema "Analyzing Various Aspects of Scheduling Independent Jobs on Identical Machines" versucht, einen im Wesentlichen theoretischen Beitrag auf dem Gebiet der **Parallel-Maschinen-Probleme** zu liefern. Dazu werden verschiedene Aspekte eines einfachen Maschinenbelegungs-Problems, bei dem n unabhängige Jobs auf m identische Maschinen so zu verteilen sind, dass eine vorgegebene Zielfunktion optimiert wird, betrachtet.

In dem zu Grunde liegenden Modell besitzt jeder Job eine nicht-negative Bearbeitungszeit, die für jede der m Maschinen gleich ist. Jeder Job muss auf genau einer der Maschinen bearbeitet werden - egal auf welcher. Dabei ist die Unterbrechung der Bearbeitung von Jobs nicht zulässig. Außerdem wird unterstellt, dass jede Maschine höchstens einen Job gleichzeitig bearbeiten kann.

Die Aufgabe besteht darin, eine Zuordnung der Jobs zu den Maschinen (genannt **Schedule**) zu finden, die eine vorgegebene Zielfunktion optimiert. Als zu optimierende Zielsetzungen werden in dieser Arbeit die Minimierung der **maximalen Maschinenlaufzeit** C_{max} (auch bekannt als **Makespan** oder Schedule-Länge), die Maximierung der **minimalen Maschinenlaufzeit** C_{min} sowie die Minimierung der Differenz $C_\Delta = C_{max} - C_{min}$ untersucht.

Da bekannt ist, dass jedes der drei Probleme NP-schwer ist, liegt das Hauptaugenmerk dieser Arbeit auf der Analyse von Heuristiken. Kapitel 2 befasst sich ausführlich mit dieser Thematik. Das vorwiegend untersuchte heuristische Lösungsverfahren ist das **List Scheduling (LS)**, welches bereits aus den Anfängen der Worst-Case Untersuchungen von Approximations-Algorithmen bekannt ist.

Beim List Scheduling liegen die n Jobs in einer beliebigen Reihenfolge als Liste (genannt **Prioritätsliste** oder **-regel**) vor, welche sukzessive abgear-

beitet wird. Sobald eine Maschine zur Bearbeitung eines Jobs frei ist, wird ihr sofort der nächste Job in der Liste zugewiesen.
Zwei der bekanntesten Prioritätsregeln sind die SPT- (**shortest processing time**) und die LPT-Regel (**longest processing time**). Bei der SPT-Regel werden die Jobs aufsteigend nach ihren Bearbeitungsdauern sortiert. Dahingegen nimmt die LPT-Regel eine absteigende Job-Sortierung vor.
Zunächst wird eine einfache obere Schranke für C_{max} und eine einfache untere Schranke für C_{min} von LS-Schedules ermittelt. Anschließend erfolgt eine Analyse der Struktur von SPT-LS-Schedules. Dabei stellt sich heraus, dass mit der SPT-Regel jeder m-te Job in der Liste der gleichen Maschine zugewiesen wird und keine anderen Jobs auf dieser Maschine bearbeitet werden. Mit diesen Ergebnissen ist es nun leicht möglich, die **Dominanz** der LPT-über die SPT-Regel für alle drei betrachteten Zielfunktionen nachzuweisen, d. h. für jede Instanz des betrachteten Problems liefert LPT einen mindestens genauso guten Zielfunktionswert wie SPT und für mindestens eine Instanz sogar einen echt besseren.
Untersuchungen hinsichtlich Dominanz-Aussagen stellen einen Schwerpunkt dieser Arbeit dar. So wird in Abschnitt 2.5 analysiert, ob es Prioritätsregeln für den LS-Algorithmus gibt, die von der SPT-Regel dominiert werden. Für alle drei Zielfunktionen kann diese Frage positiv beantwortet werden, und es wird hierfür je eine Klasse von Prioritätsregeln vorgestellt.
Eine weitere zentrale Dominanz-Aussage ist in Abschnitt 2.6 enthalten, in dem es um Verallgemeinerungen der LPT-Regel geht. Insbesondere wird die RLPT-Heuristik (**restricted longest processing time**) untersucht, die nach absteigender Jobsortierung stets m Jobs gleichzeitig so auf die m Maschinen verteilt, dass keine zwei dieser m Jobs auf der gleichen Maschine bearbeitet werden. Zunächst wird gezeigt, dass die RLPT-Heuristik die SPT-Regel hinsichtlich der drei untersuchten Zielfunktionen dominiert. Im Anschluss daran wird die RLPT-Heuristik ausführlich mit der LPT-Regel verglichen. Coffman und Sethi [CS76] zeigten bereits, dass die LPT-Regel die RLPT-Heuristik bezüglich der Zielfunktion C_{max} dominiert. Wir beweisen, dass die gleiche Aussage auch für die Zielfunktion C_{min} - und damit auch für C_Δ - gilt.
Experimentelle Untersuchungen im Zwei-Maschinen-Fall zeigen, dass die LPT-Regel in $\frac{1}{4}$ (n gerade, $n \geq 4$) bzw. $\frac{1}{8}$ (n ungerade, $n \geq 5$) der Fälle einen besseren Schedule für die drei betrachteten Zielfunktionen generiert als die RLPT-Heuristik.
Bezüglich der Worst-Case Performance kann man aussagen, dass die RLPT-Heuristik für keine Instanz einen Schedule liefert, dessen Makespan größer als das $(2 - \frac{1}{m})$-fache des optimalen Makespans ist und dessen C_{min}-Wert kleiner als ein m-tel des optimalen C_{min}-Wertes ist. Diese Performance-Schranken sind asymptotisch scharf. Es existieren aber keine Instanzen, für die diese Schranken echt angenommen werden. Die Performance-Schranken gelten in

gleicher Weise, wenn man die RLPT-Heuristik mit der LPT-Regel anstelle
von optimalem Scheduling vergleicht.

Die beiden folgenden Kapitel widmen sich Aspekten des optimalen Schedul-
ings.

In Kapitel 3 wird untersucht, welche Schedules im Zwei-Maschinen-Fall **potenziell C_{max}-optimal** sind. Ein Schedule S heißt potenziell C_{max}-optimal,
wenn es zulässige Bearbeitungszeiten der n Jobs gibt, so dass S für diese
C_{max}-optimal ist. Der Abschnitt 3.2 enthält eine vollständige Charakterisie-
rung der Menge der potenziell C_{max}-optimalen Schedules. Hierbei erweist
es sich als hilfreich, einen Schedule als eindimensionalen Pfad zu interpretie-
ren. Mit Hilfe dieser Transformation kann man nachweisen, dass jeder Schedule, dessen korrespondierender Pfad auch negative Werte aufweist, potenziell
C_{max}-optimal ist.

Abschnitt 3.3 untersucht knapp, ob Schedules auf konvexen Mengen des Po-
lytops, welches die Menge der zulässigen Instanzen beschreibt, optimal sind.
Wir nehmen hierbei ohne Beschränkung der Allgemeinheit an, dass die Be-
arbeitungszeiten aus dem Intervall $[0, 1]$ stammen. Wir zeigen anhand einer
anschaulichen Beispiel-Instanz, dass die Konvexitäts-Eigenschaft bereits im
Zwei-Maschinen-Fall nicht erfüllt ist. Im Gegensatz dazu gilt die Konvexitäts-
Eigenschaft für jede fixe Prioritätsregel und beliebige Anzahl Maschinen beim
List Scheduling.

Kapitel 4 befasst sich mit der Nicht-Äquivalenz der drei betrachteten Zielkri-
terien im Fall von mindestens drei Maschinen. Während im Zwei-Maschinen-
Fall die drei untersuchten Zielsetzungen äquivalent sind, zeigen wir anhand
minimaler Beispiel-Instanzen, dass dies bereits im Drei-Maschinen-Fall nicht
mehr gilt.

Darüber hinaus ermitteln wir experimentell im Fall von drei Maschinen die
Wahrscheinlichkeit, dass der C_{Δ}-optimale Schedule nicht C_{max}- bzw. C_{min}-
optimal ist. Die Simulationsergebnisse zeigen, dass die entsprechenden Wahr-
scheinlichkeiten monoton in der Anzahl der Jobs zunehmen und bereits für
zwölf Jobs je ungefähr $\frac{1}{10}$ betragen. Demgegenüber kann jedoch festgestellt
werden, dass in solchen Fällen die relativen Abweichungen der Zielfunk-
tionswerte C_{max} sowie C_{min} des C_{Δ}-optimalen Schedules von den optimalen
Werten selbst für kleine n fast Null sind.

Allen experimentellen Untersuchungen liegt dabei folgendes stochastisches
Modell zu Grunde: Die Bearbeitungszeiten der n Jobs sind unabhängige, auf
$[0, 1]$-gleichverteilte Zufallsgrößen.

Die vorliegende Arbeit endet mit einigen Anregungen, Ideen sowie offenen
Fragen, die eng verknüpft mit dem Inhalt dieser Dissertation sind und einen
potenziellen Ausgangspunkt für interessante zukünftige Forschungstätigkei-
ten darstellen.

Acknowledgements

The writing of this thesis has been one of the most challenging academic projects in my life. It would not have been finished in this way without the support, patience and guidance of the following people to which I like to express my deepest gratitude:

- Prof. Dr. Ingo Althöfer who acted as my supervisor for his knowledge and commitment as well as many invaluable discussions that inspired and motivated me considerably.

- My former colleagues Dr. Martin Dörnfelder, Jakob Erdmann and Lisa Schreiber who inspired my research by many helpful discussions. Dr. Martin Dörnfelder acted also as a proofreader of this thesis.

- The student Martina Vogel whose experimental results on List Scheduling inspired the work of Section 2.4.

- Prof. Dr. Armin Scholl who offered me a Post-Doc-position at his chair and gave me enough space to successfully complete this thesis.

- My parents and grandparents who have always believed in me and supported me.

Last but not least, I especially wish to thank Susanne for her affection, support and understanding. Her smile brightens my life.

Chapter 1

Introduction

Scheduling is one of the most classic topics in discrete applied mathematics and operations research. During the last decades it has been extensively studied and it is still of great academic interest. Beyond that, scheduling is of broad practical relevance. Scheduling problems in general occur in many situations. Typical applications are for instance machine scheduling in the industry, scheduling jobs on computers, timetabling at schools or universities, crew-scheduling at airlines and sequencing maintenance actions of high cost equipment such as aircraft or ships. An overview of the great variety of scheduling problems can be found in [Błe96] and [Bru95].

In our scenario, a set of independent jobs has to be scheduled without pre-emption on a set of identical machines in order to optimize some objective function. Besides production plants with parallel machines such as in the textile industry, the usage of computers with (many) processors in parallel - which is nowadays almost standard - is a good motivation for our investigations. Although precedence constraints between the execution of jobs are quite typical, we assume the jobs being independent in order to keep the problem simple concerning this matter.

This chapter introduces in detail the underlying scheduling scenario and the objective functions which are considered for it. Chapter 1 finishes with an outline of this thesis.

1.1   Problem Description

We consider the following general scheduling problem. Given are $m \geq 2$ identical parallel machines $\{M_1, \ldots, M_m\}$ and a set of $n \geq 1$ independent jobs $\mathcal{J} = \{J_1, \ldots, J_n\}$, i. e., no precedence constraints exist between any two jobs. Job J_j has a non-negative processing time of $t_j \geq 0$ units[1] and t_j is the same for all m machines. We assume the jobs to be labeled in such a way that $t_1 \geq t_2 \geq \ldots \geq t_n$. Since t_j can be zero, without loss of generality, we can assume that n is a multiple of m and t_{n-m+1} is greater than zero. Each job has to be processed by exactly one machine; regardless which one. Processing of a job is non-preemptive, i. e., once begun it must not be interrupted. Further, we assume that each machine can process at most one job at a time and each machine is available from time 0 on. The general problem is to assign the jobs to the machines so that a given objective function is optimized. The classic objective is to minimize the makespan, i. e., the time required until completion of all jobs. In other words, the makespan is the time span between the beginning of processing the first job and the completion of the last job.

A convenient notation for theoretical scheduling problems is the three field notation $\alpha|\beta|\gamma$ by Graham et al. [GLL++79]. The first field α describes the machine environment, β the job characteristics and γ the objective function. According to this, the general scheduling problem described above is denoted by:

- $\alpha = P$: identical parallel machines

- $\beta = \emptyset$: no preemption allowed, independent jobs, all jobs arrive at time 0 to the system, no deadlines

- γ needs to be specified.

Throughout this thesis we refer to it as **Identical Machine Scheduling Problem**, IMSP for short. If nothing else is mentioned, we consider the IMSP in its deterministic and static version, i. e., all input data are known beforehand. If the γ-field is specified, then a specific IMSP is given. For instance, the makespan-minimization problem is specified by $\gamma = C_{max}$. Section 1.3 introduces the objective functions that will be considered.

In the remainder of this section we provide some more terms such as rank, dummy-job and instance. We also make some agreements on how notation

---

[1]We also say that job J_j has length t_j.

and terms will be used in an economic way if no further identification is
necessary or if confusion is unlikely.

Concerning the analysis of the heuristics applied to the IMSP presented in
Chapter 2 it is useful to divide the set of jobs into $\frac{n}{m}$ **ranks**, with jobs
$J_{rm+1}, \ldots, J_{rm+m}$ in rank $r + 1$, $r = 0, \ldots, \frac{n}{m} - 1$. So, rank 1 contains the
longest m jobs of $\mathcal{J}$, rank 2 contains the longest m jobs of the remaining
jobs, etc. Finally, rank $\frac{n}{m}$ contains the shortest m jobs of $\mathcal{J}$. As described
above, to avoid redundancy we assume that at least one job in rank $\frac{n}{m}$ has
a positive processing time. We call a job with processing time equal to zero
dummy-job. For economy of notation, we omit the dummy-jobs in most of
the examples presented in this thesis. Then, n simply denotes the number of
non-dummy-jobs. In this case, n needs not to be a multiple of m.

By the term **instance** a job-system of n independent jobs (given by their
processing times) and m identical machines to be available is meant. The set
of all feasible instances is denoted by $\mathcal{I}$. When m and n need to be identified
we use (m, n)-instance and $\mathcal{I}_{m,n}$ instead.

For the sake of simplicity we often identify both, jobs and machines, by their
index.

1.2   Solution Representation

A feasible assignment of the jobs to the machines is called **schedule**. Usually
a schedule contains not only information about which job is processed by
which machine but also starting and finishing times of the jobs.

For the questions studied in this thesis it is of minor interest in which order
jobs are processed by the machine they are assigned to. Moreover, due to the
fact that we consider identical machines it is only interesting which jobs are
assigned to the same machine, but not to which one. Hence, in our context a
schedule is simply a partition of $\mathcal{J}$ into (at most) m subsets. But with regard
to Chapter 3 we use another way of representing schedules instead, namely
strings of length n over the set $\{1, 2, \ldots, m\}$. Of course, this representation is
not unique. However, this can be corrected by considering only non-permuted
schedules which are defined as follows.

Definition 1.2.1 (Non-permuted schedule)
*A schedule $S \in \{1, 2, \ldots, m\}^n$ is called non-permuted if S fulfills the following
two conditions:*

(*i*) $S(1) = 1$

(*ii*) $S(j) \in \left\{ 1, \ldots, \min\{m, 1 + \max_{1 \leq k \leq j-1} S(k)\} \right\}$ *for all* $j = 2, \ldots, n$.

The understanding of $S(j) = i$ is, that job j is assigned to machine i in schedule S.

Let $S_{n,i}$ be the Stirling number of second kind, which counts the number of ways to partition a set of n elements into i non-empty subsets [Ste02, page 32]. Then, for given m and n, $\sum_{i=1}^{m} S_{n,i}$ gives the total number of non-permuted schedules because each non-permuted schedule belongs bijectively to one partition of $\mathcal{J}$. In [MR05] an explicit formula of $\sum_{i=1}^{m} S_{n,i}$ as a linear combination of i^n for $i \in \{1, \ldots, m\}$ is determined. It is also shown that for each $m \geq 1$ the ratio $\frac{\sum_{i=1}^{m} S_{n,i}}{m^n}$ converges to $\frac{1}{m!}$ if n approaches infinity. So, for fixed m and large n the number of non-permuted schedules is approximately $\frac{m^n}{m!}$.

As mentioned above, we consider only non-permuted schedules in this thesis. For the sake of simplicity we usually omit the word *non-permuted* in the following.

1.3 Objective Functions and Preliminary Results

This section presents and motivates the three objective functions for the IMSP that are considered in this thesis. Preliminary results such as trivial bounds on the optimal objective function values and complexity results are also contained.

Objective Functions

At first, a new term is introduced. For a given schedule S, C_i denotes the **completion time** of machine i. Since precedence constraints do not exist between any two jobs and the machines are always available, idle times between finishing a job and starting another one on the same machine do not occur. So, the completion time of a machine is simply the sum of the processing times of all jobs assigned to that machine. More formally, $C_i = \sum_{j:S(j)=i} t_j$ for all machines i. When the schedule S needs to be identified we use C_i^S.

We consider the following three objective functions which lead to specific IMSP's:

- maximum completion time, denoted by $C_{max} = \max_{i=1,\ldots,m} C_i$, which is to be minimized (C_{max} is also called **makespan** or **schedule length**)

- minimum completion time, denoted by $C_{min} = \min_{i=1,\ldots,m} C_i$, which is to be maximized

- difference between maximum and minimum completion time, denoted by $C_\Delta = C_{max} - C_{min}$, which is to be minimized.

According to the three field notation introduced in Section 1.1 we choose γ from the set $\{C_{max}, C_{min}, C_\Delta\}$.

Makespan-minimization, i. e., $\gamma = C_{max}$, is the most classic objective for scheduling problems at all. It belongs to the class of packing problems as the jobs should be "packed" into the smallest possible time interval on all machines. Practical motivations for this objective are the wish for schedules that yield high throughput or high machine utilization which is implied by short makespan-schedules.

For the analyses presented in the next chapters it is comfortable to introduce a new term.

Definition 1.3.1 (Makespan-Machine)
*Let S be a schedule. A machine i is called **makespan-machine**, if $C_i^S = C_{max}^S$.*

The problem of maximizing C_{min}, i. e., $\gamma = C_{min}$, belongs to the class of covering problems as the jobs should "cover" the longest possible time interval that is common to all machines. This situation can be motivated by the following scenario. A system is alive (i. e. productive) only if all of its machines are alive. The machines (which are identical) consist of one main part that needs to be replaced after a certain period of time. Concerning the replacements, a finite number of spare parts is available. Each spare part has a known (deterministic) life-time and is compatible with all machines. Each spare part can be used exactly one time. The duration a machine is alive is given by the sum of life-times of all spare parts used for that machine. The goal is to keep the system alive as long as possible. This problem has applications in the sequencing of maintenance actions for modular gas turbine aircraft engines [FD81]. Considering the spare parts as jobs leads us back to the IMSP with objective function C_{min}.

The problem of minimizing C_Δ, i. e., $\gamma = C_\Delta$, comprises both C_{max}-minimization and C_{min}-maximization. Thus, it seems to be the most direct measure of "near-equality"[2]. This problem can be motivated by the wish for schedules that utilize the available machines almost equally. If the machines are run by operators, C_Δ ensures that the total work is distributed almost equally among the workers.

In literature [CL84], these three problems are also known as specific partitioning problems which are classic in combinatorial optimization. In particular, the makespan-minimization problem is known as number partitioning and the C_Δ-minimization problem is known as set partitioning.

Obviously, the three objective functions seem to be closely related as they can be seen as measures of "near-equality". In case of two machines they are completely equivalent, because here the equality $C_{min} = \sum_{i=1}^{n} t_i - C_{max}$ holds. However, in systems with at least three identical machines we present in Chapter 4 small instances for which the optimal schedule concerning one of the three objective functions is not optimal for at least one of the other two objective functions.

Both C_{max} and C_{min}, but particularly C_{max}, are of major interest in this work. We often use results for these two objective functions to derive statements concerning C_Δ.

Preliminary Results

If we assume the processing times to be positive integers and $m = 2$, then the decision problem whether a perfect partition (i. e. a schedule for which C_Δ is zero - or one if the sum of all processing times is odd) exists is known to be NP-complete [Kar72]. This problem is also called PARTITION. The corresponding optimization problem is NP-hard.

To show NP-hardness of each of the three specific IMSP's defined before is straightforward in case of more than two machines. For instance, this can be done by application of the restriction technique [GJ79] to the corresponding decision problem.

To complete this section we give some obvious bounds on the optimal objective function values. Throughout this thesis, the superscript * usually

---

[2]The term "near-equality" has already been used in [CL84].

refers to optimal scheduling. It indicates the optimal objective function value or a corresponding optimal solution. For clarification, $S^*_{C_{max}}(I)$ denotes a C_{max}-optimal schedule, i. e., an optimal solution to the C_{max}-minimization problem, concerning instance $I \in \mathcal{I}$. The optimal makespan is denoted by $C^*_{max}(I)$. If no identification of I is needed, $S^*_{C_{max}}$ represents a C_{max}-optimal schedule and C^*_{max} is used for the optimal makespan. In an analogous manner we use these notation for the C_{min}-maximization problem as well as the C_Δ-minimization problem.

Lemma 1.3.2
$C^*_{max} \geq \max\{t_1, \frac{1}{m}\sum_{j=1}^n t_j\}$.

Proof
Some machine has to process the longest job, i. e., $C^*_{max} \geq t_1$.
On the other hand, the total work is $\sum_{j=1}^n t_j$. So, at least one machine must have a completion time of at least a $\frac{1}{m}$-fraction of the total work, i. e., $C^*_{max} \geq \frac{1}{m}\sum_{j=1}^n t_j$. Hence, $C^*_{max} \geq \max\{t_1, \frac{1}{m}\sum_{j=1}^n t_j\}$. ∎

Lemma 1.3.3
$C^*_{min} \leq \frac{1}{m}\sum_{j=1}^n t_j$.

Proof
At least one machine must have a completion time of at most a $\frac{1}{m}$-fraction of the total work, i. e., $C^*_{min} \leq \frac{1}{m}\sum_{j=1}^n t_j$. ∎

Lemma 1.3.4
$C^*_\Delta \geq 0$.

Proof
The proof is clear by the definition of C_Δ. ∎

It is readily verified that $C^*_{max} > 0 \Leftrightarrow n \geq 1$ and $C^*_{min} > 0 \Leftrightarrow n \geq m$ whereby n denotes the number of non-dummy-jobs here. Whenever we consider worst-case bounds concerning C_{min}, we assume that the job-systems consist of at least m non-dummy-jobs. Hence, these two objective functions are appropriate measures to derive meaningful worst-case bounds on the performance of heuristics relative to an exact algorithm.
This is not the case for C_Δ. Here, it is quite simple to construct (m,n)-instances so that $C^*_\Delta = 0$ for $n \geq m$ and any fixed m. Consequently, C_Δ is not an appropriate measure to derive meaningful worst-case bounds on the performance of heuristics relative to an exact algorithm. Therefore, instead

of minimizing C_Δ one might minimize the ratio $\frac{C_{max}}{C_{min}}$ which has already been studied in [CL84]. However, for two reasons we still keep to the measure C_Δ. Firstly, in our analysis of different heuristics presented in Chapter 2 we do not intend to derive worst-case bounds on the C_Δ-performance of the heuristics. Secondly, C_Δ seems to be the most direct and intuitive measure of "near-equality" which makes it more interesting for our investigations concerning optimal scheduling presented in Chapter 4.

1.4 Example

This section is meant to repeat and illustrate the main terms of the previous sections by means of a concrete instance of the IMSP. The reader who is familiar with scheduling terminology can easily skip this section.

We consider the following $(3, 6)$-instance I given by the vector of processing times $I = (18, 12, 11, 9, 8, 6)$. For economy of notation, the symbol I will denote both the vector of processing times and the name of the instance itself. Instance I consists of $m = 3$ (identical) machines and $n = 6$ jobs, which are subdivided into two ranks. Rank 1 contains the three longest jobs, i. e., the corresponding processing times in rank 1 are $18, 12$ and 11. Rank 2 contains the three shortest jobs, i. e., the corresponding processing times are $9, 8$ and 6. We write $R_1 = (18, 12, 11)$ and $R_2 = (9, 8, 6)$ for short. Instance I does not contain dummy-jobs since all six processing times are positive.

A schedule for I is for instance $S_1 = (1, 2, 2, 3, 3, 3)$. According to S_1, the longest job is processed by machine 1, the second and the third longest job are processed by machine 2 and the three shortest jobs are processed by machine 3. The partition $\{\{18\}, \{12, 11\}, \{9, 8, 6\}\}$ of the processing times corresponds to S_1. Once again, note that in a given schedule it is actually sufficient to know which jobs are mutually processed by a machine, but not on which machine. Despite this fact we keep to the string-representation in order to have a unified representation of schedules. Besides S_1 we do also consider the schedules $S_2 = (1, 2, 3, 3, 2, 1)$ and $S_3 = (1, 2, 3, 1, 2, 3)$. All of these three schedules are non-permuted. The total number of non-permuted schedules for $(3, 6)$-instances is $\frac{1}{2}(3^5 + 1) = 122$.

Table 1.1 on page 9 contains the machine completion times and the objective function values of the three schedules S_1, S_2, S_3.

In the schedule S_1 the machines 2 and 3 are makespan-machines. In the schedules S_2 and S_3 machine 1 is the makespan-machine.

	C_1	C_2	C_3	C_{max}	C_{min}	C_Δ
$S_1 = (1, 2, 2, 3, 3, 3)$	18	23	23	**23**	18	5
$S_2 = (1, 2, 3, 3, 2, 1)$	24	20	20	24	**20**	**4**
$S_3 = (1, 2, 3, 1, 2, 3)$	27	20	17	27	17	10

Table 1.1: Machine completion times and objective function values of S_1, S_2 and S_3 for I

Schedule S_1 is an optimal assignment for the makespan-minimization problem, i. e., $C^*_{max}(I) = 23$, whereas S_2 is optimal for both C_{min}-maximization and C_Δ-minimization, i. e., $C^*_{min}(I) = 20$ and $C^*_\Delta(I) = 4$. Considering only non-permuted schedules as it is done in this thesis, it is easy to see that S_1 is the unique solution to the C_{max}-minimization problem, and S_2 is the unique solution to the other two specific IMSP's for I. In comparison to this, Schedule S_3 performs rather poorly for any of the three objective functions. However, in Chapter 2 we will see that one of the heuristics analyzed there generates S_3 for I.

To sum up, this small instance I is an example for the fact that the (unique) solution to the C_{max}-minimization problem neither needs to be optimal for C_{min} nor for C_Δ, even in case of only three machines. We refer to Chapter 4 where we present more instances concerning this and related issues.

1.5 Outline of this Thesis

This introductory chapter finishes with a survey of how the remainder of this thesis is structured.

The main attention of this work is given to the analysis of the List Scheduling (LS) algorithm. This approximation algorithm is extensively studied in Chapter 2. It starts with an introduction to List Scheduling and two of the most popular priority rules for it, the SPT-rule (shortest processing time) and the LPT-rule (longest processing time).

Before the main results of our investigations concerning List Scheduling are presented, we give a brief literature review in Section 2.2. Afterwards in Section 2.3, we derive a simple bound on the maximum as well as the minimum completion time of LS-schedules. The next two sections are dedicated to the SPT-rule. In Section 2.4, we analyze the structure of SPT-LS-schedules, and we use these insights to compare the SPT- and the LPT-rule concerning their objective function values. Section 2.5 contributes to the question whether the

SPT-rule is the worst priority rule of the LS-algorithm. This question will be studied in detail for all three objective functions C_{max}, C_{min} and C_Δ. The section finishes with an average-case analysis of the SPT-rule. We describe the underlying stochastic model which is used throughout this thesis whenever we present experimental results (unless anything else is mentioned). Some experimental results concerning the expected makespan of LPT-scheduling in case of two machines are also contained.

In Section 2.6 we investigate LPT-based heuristics. In particular, we take a detailed look at the RLPT-heuristic (restricted longest processing time). We compare the RLPT-heuristic to SPT- and LPT-scheduling. Furthermore, we study the worst-case performance of the RLPT-heuristic relative to optimal scheduling as well as to LPT-scheduling concerning the two problems C_{max}-minimization and C_{min}-maximization. We also present experimental results concerning our comparison of the RLPT- and the LPT-heuristic in case of two machines. The section finishes with a generalization of both RLPT- and LPT-scheduling.

The next two chapters deal with optimal scheduling. In Chapter 3 we investigate the question which schedules are potentially C_{max}-optimal in case of two machines. Therefore, in Section 3.1 we introduce a transformation which allows to interpret a schedule as a one-dimensional path. With this transformation we are able to give a total characterization of potentially C_{max}-optimal schedules in case of two machines in Section 3.2. The chapter ends in Section 3.3 with a result concerning convexity.

Chapter 4 briefly contributes to the non-equivalence of the three objective functions $C_{max}, C_{min}, C_\Delta$ in case of at least three machines by presenting smallest instances. We compare each pair of objective functions, and we also conduct a joint comparison of all three objective functions. Finally, we present some experimental results on the probability that the C_Δ-optimal schedule is not C_{max}- or C_{min}-optimal in case of three machines.

In Chapter 5 we summarize the main results of this thesis, and we discuss ideas as well as open questions for future work.

After having read Chapter 1 which introduces the basic terms and notation, the Chapters 2, 3 and 4 can be read (almost) independently from each other. This means that it is recommended to read Subsection 2.1.1 before reading Section 3.3. Furthermore, it is also advisable to read Subsection 2.5.2 before reading Subsection 2.6.3.3 or Section 4.6 in order to be informed about the underlying stochastic model.

Regarding Chapter 2, the Sections 2.5 and 2.6 can be read independently from each other after the previous Sections 2.1, 2.3 and 2.4 have been read in this order.
The general hints on how to read this thesis are illustrated by the following figure.

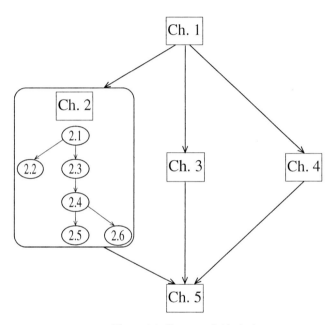

Figure 1.1: How to read this thesis

Chapter 2

Heuristics

In Section 1.3, we stated that each of the three specific IMSP's is NP-hard. Due to this fact it seems unlikely that efficient algorithms will be found for obtaining optimal solutions to these problems. For this reason much effort has been and is still put in the development and analysis of different heuristics.

In contrast to exact algorithms, heuristics aim at finding good or near-optimal solutions in a reasonable amount of time, generally using quite intuitive and promising techniques. When designing a heuristic, common demands are to run in polynomial time, to solve arbitrary instances of the underlying problem, and to have a (meaningful) guarantee on the worst-case performance of the heuristic relative to an exact algorithm. If a heuristic applied to a specific problem fulfills all of these three features we typically use the term **approximation algorithm** for it [Bla96], [Hal97] and [KT05]. In addition, the average-case performance of a heuristic solution procedure also plays a crucial role.

Historically, scheduling problems were among the first problems to be analyzed with respect to the worst-case behavior of approximation algorithms. It probably started in 1966, when Ronald Graham presented a worst-case analysis of a simple algorithm known as List Scheduling [Gra66]. This chapter is dedicated to this algorithm and to generalizations of the LPT-rule (longest processing time), which is one of the most famous priority rules for the List Scheduling algorithm.

Priority rules [DSV97],[Hau89] are simple procedures that specify in which order jobs are selected for processing if machines become available. In more complex scheduling problems such as shop scheduling problems where each job consists of several tasks that have to be processed by certain machines, priority rules are often applied to generate initial solutions. Then, these solutions are typically used as starting points of local search heuristics such as Simulated Annealing, Tabu Search and Threshold Accepting [AL97], [Bła96]. Local search heuristics aim at a further improvement of feasible solutions by local exchanges. The job shop scheduling problem and the open shop scheduling problem are well-known representatives of shop scheduling problems.

We begin this chapter by describing the List Scheduling algorithm, present its worst-case performance and introduce two popular priority rules, namely SPT (shortest processing time) and LPT, that we will compare. A brief overview of related work is given in the second section. The next sections include a comparison of the objective function values produced by the SPT- and the LPT-rule. Moreover, we show that priority rules exist for the List Scheduling algorithm that are dominated by the SPT-rule. Finally, Section 2.6 deals with LPT-based heuristics and their analysis.

2.1   List Scheduling

This section deals with a simple algorithm, known as **List Scheduling** (LS), from the beginnings of worst-case analyses of approximation algorithms.

2.1.1   The List Scheduling Algorithm

Concerning the (general) IMSP defined in Section 1.1, LS works as follows:

1. The n jobs are given in a list L in some arbitrary order.

2. Whenever a machine completes a job, it immediately begins processing the first job in the list which has not yet begun to be executed. Tie situations, in which two or more machines attempt to begin the same job, are broken arbitrarily.

The list L is usually called job list or **priority list** indicating the relation to priority rules. We understand L as a permutation of $\mathcal{J}$. $L(k) = j$ means that job j, i. e., the j-th longest job, appears at position k in L. Note that $L(k) = j$ is equivalent to $L^{-1}(j) = k$.

Since we are rather interested in objective function values than in corresponding schedules in this chapter, we make the following agreement concerning tie situations in step 2 of LS. If nothing else is mentioned, it will be our convention to assign a job in a tie situation to the machine with the smallest index out of the corresponding machines. With this convention, LS becomes a deterministic algorithm. Note that this tie-break rule does generally not produce non-permuted schedules. However, this is not a problem as we can easily transfer LS-schedules into non-permuted ones.

We also remark that the deterministic version of LS justifies to speak of *the* LS-schedule instead of *a* LS-schedule for a given priority list L (and a job-system). Furthermore, the deterministic version does not influence the values of C_{max}, C_{min} and C_Δ.

Of course, LS can be implemented to run in polynomial time. Using a priority queue, LS runs in $\mathcal{O}(n \log m)$ time.

Why seems LS to be an appropriate algorithm concerning the three specific IMSP's of this thesis? Step 2 of LS assigns the next job on the list to a machine with least completion time so far whenever a machine completes a job. Considered locally, i. e., given a partial schedule and only the next job to be scheduled, this decision seems to be meaningful since it ensures that

- the current C_{max}-value increases as little as possible and

- the current C_{min}-value increases as much as possible.

Hence, it also yields the best current C_Δ-value that is locally achievable. Furthermore, in [CG85] it is mentioned that for each instance of the make-span-minimization problem a priority list exists which leads to a C_{max}-optimal schedule applying LS. Of course, finding the corresponding order is NP-hard. Considering the example instance $I = (18, 12, 11, 9, 8, 6)$ of Section 1.4, the priority list $L = (1, 2, 4, 5, 3, 6)$ generates the C_{max}-optimal schedule $S = (1, 2, 2, 3, 3, 3)$. Analogously, priority lists exist which lead to optimal schedules concerning C_{min} and C_Δ, respectively. The list $L = (1, 2, 3, 4, 5, 6)$ generates the C_{min}- and C_Δ-optimal schedule $S = (1, 2, 3, 3, 2, 1)$ for I.

2.1.2 Worst-Case Analysis

The LS-algorithm is probably the first approximation algorithm for which a worst-case analysis has been presented [Gra66]. In 1966, Ronald Graham showed that the makespan of the LS-schedule, C_{max}^{LS}, is never more than $(2 - \frac{1}{m})$ times the optimal makespan C_{max}^*. We say that the **performance bound** of LS is $(2 - \frac{1}{m})$ concerning makespan-minimization. If it is clear from

the context, we do not explicitly add to which objective the performance
bound is related.
Since Graham's result is such a basic one to which we relate a few times in
this chapter, we repeat the proof here.

Theorem 2.1.1 (Graham, 1966)
An LS-schedule is at most $(2-\frac{1}{m})$ *times as long as the optimal schedule, i. e.,*
$C_{max}^{LS} \leq (2-\frac{1}{m})C_{max}^*$. *Moreover, this bound is tight for any fixed number* m
of machines.

Proof
Consider an arbitrary instance of the IMSP and an arbitrary priority list.
Let machine i be a makespan-machine of the LS-schedule, i. e., $C_{max}^{LS} = C_i^{LS}$.
Furthermore, let j be the last job that is processed by machine i in the
LS-schedule. When job j is assigned to machine i, this machine had the
smallest completion time so far. Its completion time before assigning job j
is $(C_i^{LS} - t_j)$ which is not greater than any C_k^{LS} for $k = 1, \ldots, m$. Summing
up all inequalities leads to:

$$C_i^{LS} - t_j \leq C_k^{LS} \quad (k = 1, \ldots, i-1, i+1, \ldots, m)$$

$$\Rightarrow (m-1)(C_i^{LS} - t_j) \leq \sum_{\substack{k=1 \\ k \neq i}}^{m} C_k^{LS}$$

$$\Rightarrow (m-1)(C_i^{LS} - t_j) + C_i^{LS} \leq \sum_{k=1}^{m} C_k^{LS} = \sum_{h=1}^{n} t_h$$

$$\Rightarrow mC_i^{LS} \leq \sum_{h=1}^{n} t_h + (m-1)t_j \leq \sum_{h=1}^{n} t_h + (m-1)t_1$$

$$\Rightarrow C_i^{LS} \leq \frac{1}{m}\sum_{h=1}^{n} t_h + \frac{m-1}{m}t_1$$

$$\leq C_{max}^* + \frac{m-1}{m}C_{max}^* = (2 - \frac{1}{m})C_{max}^*.$$

The last inequality holds due to Lemma 1.3.2.

To verify that this bound is tight consider m machines, $m(m-1)$ jobs of
length 1 and one job of length m. Obviously, the optimal makespan is m
by processing the one job of length m on a separate machine and each of
the remaining $(m-1)$ machines processes m jobs of length 1. Then, each
machine has completion time m. Now, consider LS for such instances and
assume the one length m job is the last job in the list. After the assignment

of the first $m(m-1)$ jobs of the list, each machine runs $(m-1)$ time units in the LS-schedule. One of the m machines must process the last job of the list. This machine has completion time $(2m-1)$, whereas all other machines have completion time $(m-1)$. Hence, the performance bound of LS concerning C_{max}-minimization is tight. ∎

If a performance bound is tight, we usually call it **performance ratio** instead of performance bound. That is, LS has a performance ratio of $(2 - \frac{1}{m})$. In literature, the terms $(2 - \frac{1}{m})$-approximation or asymptotically 2-approximation are used, too.

The worst-case analysis of LS compared to optimal scheduling for C_{max} presented above does not make any special assumptions on the order of the jobs in L. Hence, applying LS in an online scenario where jobs may arrive over time to the system, the makespan of the LS-schedule is also guaranteed to not exceed $(2 - \frac{1}{m})$ times the optimal makespan. However, in this thesis we keep to static scenarios where all input data are known beforehand.

Concerning the problem of maximizing C_{min}, Woeginger [Woe97] proved the following tight lower bound on the performance ratio of LS compared to optimal scheduling:

$$\frac{C_{min}^{LS}}{C_{min}^{*}} \geq \frac{1}{m}.$$

To verify that this bound is tight for any fixed number m of machines (Woeginger did not give an instance in [Woe97]), we present the following $(m, 2m - 1)$-instance:

$$t_1 = \ldots = t_{m-1} = m,$$
$$t_m = \ldots = t_{2m-1} = 1.$$

Then, the C_{min}-optimal schedule is obtained by processing the m shortest jobs on the same machine and each of the $(m-1)$ longest jobs on a separate machine. Hence, each machine has completion time m. Now, consider a priority list L of the LS-algorithm so that the m shortest jobs are assigned first, e. g., $L = (m, m+1, \ldots, 2m-1, 1, 2, \ldots, m-1)$. This leads to an LS-schedule where $(m-1)$ machines have completion time $(m+1)$ and one machine has completion time 1. Thus, the bound is tight.

Deriving a meaningful bound on the performance ratio of a heuristic relative to an exact algorithm concerning the objective function C_Δ is not appropriate as mentioned in Section 1.3.

2.1.3   Two Special LS-rules

If the jobs are ordered in a pre-specified way, we speak of an **LS-rule** or **LS-heuristic**. In this thesis, we concentrate on two famous LS-rules:

(i) The SPT-rule sorts the jobs in non-decreasing order according to the processing times, i. e., $L = (n, n-1, \ldots, 2, 1)$. SPT stands for **shortest processing time**.

(ii) The LPT-rule sorts the jobs in non-increasing order according to the processing times, i. e., $L = (1, 2, \ldots, n)$. LPT stands for **longest processing time**.

Each of the two LS-rules has a time complexity of $\mathcal{O}(n \log n + n \log m)$, $\mathcal{O}(n \log n)$ being the effort for sorting the jobs.

Since each of the two heuristics needs to know the processing times of all jobs to be scheduled beforehand, they are usually called offline-heuristics. Motivation for the analysis of these two special heuristics is included in Section 2.4. If confusion is unlikely we refer to these two rules as SPT and LPT instead of using SPT-LS and LPT-LS, respectively.

Applying these two rules to instance $I = (18, 12, 11, 9, 8, 6)$ from Section 1.4, the SPT-rule generates the schedule $S^{SPT} = (1, 2, 3, 1, 2, 3)$, whereas LPT generates $S^{LPT} = (1, 2, 3, 3, 2, 1)$. Using the schedule notation from 1.4, S^{SPT} equals the rather poor performing schedule S_3 and S^{LPT} equals the C_{min}- and C_Δ-optimal schedule S_2.

To sum up, the SPT-rule applied to instance I performs worse than the LPT-rule concerning all three objectives. We generalize this result by Theorem 2.4.3 on page 28.

Before we start with our analysis of the SPT- and LPT-rule we want to briefly overview related work on LS and other heuristic approaches to the IMSP.

2.2   Related Work

Because of the large amount of related work, this section is not meant to give an exhaustive review. We rather restrict to the work that is in our opinion most interesting for the content of this chapter. However, we still try to cover the main (algorithmic) approaches to the IMSP. At appropriate points in the remainder of this chapter we add some more results from literature. The overview on related work starts with important results on the worst-case as well as the average-case performance of the LPT-rule. Afterwards, we briefly introduce and discuss the Multifit algorithm and the Differencing Method which are competitive to LPT-scheduling.

As already mentioned, in 1966 Ronald Graham presented the probably first worst-case analysis of an approximation algorithm proving the makespan of the LS-schedule to be never more than $(2 - \frac{1}{m})$ times the optimal makespan [Gra66]. Three years later, in 1969 Graham presented a more sophisticated worst-case analysis of the LPT-rule [Gra69]. There, he proved that the makespan of the LPT-schedule is at most $(\frac{4}{3} - \frac{1}{3m})$ times the optimal makespan which is again a tight bound for any fixed number m of machines.

The worst-case performance of the LPT-rule for the C_{min}-maximization problem has also been studied successfully. Deuermeyer, Friesen, and Langston [DFL82] showed that the *minimum* completion time of the LPT-schedule is never less than $\frac{3}{4}$ times the optimal *minimum* completion time. This bound is asymptotically tight as m approaches infinity. The tight performance bound of LPT for the problem of maximizing C_{min} is presented by Csirik, Kellerer and Woeginger [CKW92]. They proved that the minimum completion time of the LPT-schedule is at least $\frac{3m-1}{4m-2}$ times the optimal minimum completion time. This bound is tight for any fixed m.

Besides worst-case analyses (based on combinatorial models), a growing interest in the expected or average-case performance (based on distribution models) of approximation algorithms for scheduling problems arose in the eighties of the last century.

Coffman and Gilbert [CG85] studied the expected relative performance of List Scheduling. They obtained bounds on the expected ratio $\frac{C_{max}^{LS}}{C_{max}^*}$. The processing times are treated as independent identically distributed random variables which is commonly done in average-case analyses.

For LPT, probabilistic analyses are given by Frenk and Rinnooy Kan [FR86], [FR87] confirming good asymptotic properties of this priority rule in a strong sense (absolute rather than relative error). In [FR87] it is shown that (under mild conditions on the distribution of the processing times) the absolute error $\left(C_{max}^{LPT} - C_{max}^*\right)$ converges to 0 almost surely as well as in expectation when n tends to infinity. The authors also analyzed the speed at which convergence to absolute optimality occurs. They proved that if the processing times are independent uniformly distributed on $[0, 1]$, then the absolute error $\left(C_{max}^{LPT} - C_{max}^*\right)$ is at most $\mathcal{O}(\frac{\log n}{n})$ almost surely and $\mathcal{O}(\frac{1}{n})$ in expectation. For almost sure convergence, the speed result also holds in case of independent exponentially distributed processing times. An extension and generalization of the results on almost sure convergence and convergence in expectation to absolute optimality of the LPT-rule is contained in [FR86][1].

In this context we also mention the contribution of Rinnooy Kan [Rin86] which provides a good introduction to the analysis of approximation algorithms. Concerning the makespan-minimization problem, Rinnooy Kan considers the LS-heuristics SPT and LPT as well as the Differencing Method in brief. Different approaches to the analysis of heuristics are presented and motivated, too. We also advise the book by Coffman and Lueker [CL91] which is concerned with a wide variety of techniques concerning the probabilistic analysis of algorithms for partitioning problems.

In case of two identical machines, Coffman, Frederickson, and Lueker [CFL84] analyzed the expected makespans of LPT and RLPT. With the RLPT-rule (restricted longest processing time), jobs are assigned in pairs, one to each machine[2]. The authors assume the processing times to be independent samples from the uniform distribution on $[0, 1]$. They proved the following results: $\frac{n}{4} + \frac{1}{4(n+1)} < \mathbb{E}\left[C_{max}^{LPT}\right] < \frac{n}{4} + \frac{e}{2(n+1)}$ and $\mathbb{E}\left[C_{max}^{RLPT}\right] = \frac{n}{4} + \frac{1}{2(n+1)}$. We refer to these results in Section 2.6.3.3 where we investigate some related questions. For $m = 2$, Tsai [Tsa92] proposed an algorithm that produces cardinality-balanced schedules for which C_Δ is bounded by $\mathcal{O}(\frac{\log n}{n^2})$ almost surely. In a cardinality-balanced schedule, the number of non-dummy-jobs assigned to each machine is either $\lceil \frac{n}{m} \rceil$ or $\lfloor \frac{n}{m} \rfloor$. Note that RLPT produces cardinality-balanced schedules, too.

Comparing LPT- and RLPT-schedules, Coffman and Sethi [CS76] proved that for each (m, n)-instance the makespan of the LPT-schedule is not worse than the makespan of the RLPT-schedule, i. e., $C_{max}^{LPT} \leq C_{max}^{RLPT}$. Although (non-trivial) results concerning the dominance of one heuristic over another one are quite interesting as they allow for eliminating dominated heuristics, they seem to be rather rare. Therefore, one goal of this thesis is to contribute to this issue.

An alternative algorithm is Multifit which was proposed by Coffman, Garey, and Johnson [CGJ78]. The Multifit algorithm is based on techniques from bin-packing and has a better worst-case performance than the LPT-rule in terms of makespan-minimization. Multifit starts with an upper bound C_{upper} and a lower bound C_{lower} for the makespan and then applies a binary search (over k iterations). At each iteration, the bin-packing heuristic First-Fit

---

[1]Note that the paper [FR86] appeared earlier than [FR87] although it had been submitted later. The authors refer in [FR86] to their paper [FR87].

[2]For more details concerning RLPT we refer to Section 2.6.

Decreasing (FFD) is applied with capacity $C = \frac{1}{2}(C_{lower} + C_{upper})$, where FFD sorts the jobs in non-increasing order and then sequentially packs each job in the lowest numbered bin in which it fits. If FFD finds a feasible solution, i. e., all n jobs can be assigned to the m machines constrained to the maximal availability of C time-units of each machine, the upper bound is set to $C_{upper} := C$. Otherwise, the lower bound is set to $C_{lower} := C$. This is iterated k times. The final C_{upper} is the makespan obtained by Multifit. The total time complexity of Multifit is $\mathcal{O}(n \log n + kn \log m)$.

Coffman, Garey, and Johnson [CGJ78] proved that the performance ratio of Multifit concerning makespan-minimization is $\frac{8}{7}$ for $m = 2$, $\frac{15}{13}$ for $m = 3$, and $\frac{20}{17}$ for $m = 4, 5, 6, 7$. For all $m > 7$, Yue [Yue90] proved that $\frac{13}{11}$ is an upper bound on the performance ratio of Multifit. A few years earlier, Friesen [Fri84] had already presented an instance which shows that the $\frac{13}{11}$-bound is tight for all $m \geq 13$.

By modifying[3] Multifit, Friesen and Langston [FL86] obtained an algorithm for which $\frac{72}{61}$ is an upper bound on the performance ratio for C_{max}-minimization. They also showed that this bound is tight for all $m \geq 12$.

Although Multifit provides a better worst-case performance, in many cases LPT still generates better schedules concerning C_{max}. This motivated Lee and Massey [LM88] to develop a combined algorithm which uses the output of LPT as the initial upper bound for Multifit which is then applied with fewer iterations. Hence, the makespan of the schedule constructed by this combined algorithm is not worse than the makespan of the LPT-schedule. Lee and Massey also proved that the performance bound of the combined algorithm with $(k - 1)$ iterations of FFD is not worse than that of Multifit with k iterations of FFD. Furthermore, for $m = 2$ they proved that the performance ratio of the combined algorithm is $\frac{10}{9}$. This is better than that of the constituent two algorithms.

The last general algorithm that we want to present in this section is the Differencing Method which was proposed by Karmarkar and Karp [KK82] in the year 1982 for the C_Λ-minimization problem. This algorithm outperforms any other known (practical) polynomial time algorithm from an average-case perspective.

We want to explain the idea of the Differencing Method for the special case of two machines because it is most illustrative for it. The key ingredient is the differencing operation: starting with a list of all n numbers, i. e., the processing times, at each iteration two numbers of the list are selected and

---

[3]The modification concerns the case in which FFD is not capable of finding a feasible solution for the current capacity C.

replaced by their absolute difference. After $(n-1)$ iterations only one number is left, which is the final C_Δ-value. The replacement of two numbers in the list by their absolute difference means assigning them into different subsets, i. e., to different machines, while deferring the decision to which subset each number will be assigned. By backtracking through the successive differencing operations, one can easily determine the corresponding two subsets.

Several strategies have been proposed for selecting the pairs of numbers to be differenced. For instance, the Largest Differencing Method (LDM) always selects the largest two numbers in the list for differencing. For $m = 2$, Yakir [Yak96] proved that if the processing times are independent uniformly distributed on $[0, 1]$, then the expected value of C_Δ produced by LDM ($\mathbb{E}[C_\Delta^{LDM}]$) is asymptotically $n^{-\Theta(\log n)}$. A recent result concerning this issue is due to Boettcher and Mertens [BM08] who contributed to the constants in the Θ-term. More precisely, Boettcher and Mertens argue that $\mathbb{E}[C_\Delta^{LDM}] = n^{-c \ln n}$, where $c = \frac{1}{2 \ln 2}$.

An alternative strategy, called Paired Differencing Method (PDM), is due to Lueker [Lue87] and works as follows. In each phase of PDM the numbers in the list are sorted non-increasingly and the largest two numbers, the third and fourth largest number, etc., are paired for differencing. The phases are iterated until only one number is left. Note that each phase of PDM halves the length of the list and that the schedule generated by PDM is cardinality-balanced. In case of independent uniformly distributed processing times on $[0, 1]$ and $m = 2$, Lueker [Lue87] showed that the expected C_Δ-value of PDM is $\Theta(\frac{1}{n})$.

By combining PDM and LDM to the Balanced Largest Differencing Method (BLDM) for $m = 2$, Yakir [Yak96] proposed an algorithm that produces cardinality-balanced schedules, and whose average-case performance is comparable to that of LDM. In this context, Mertens [Mer99] proposed a complete anytime algorithm that optimally solves the problem of finding a shortest cardinality-balanced schedule in case of two machines.

For $m \geq 2$, Karmarkar and Karp [KK82] presented a rather sophisticated differencing method that does not necessarily generate a cardinality-balanced schedule. They used some randomization in selecting the pair that is to be differenced and proved that the expected C_Δ-value of their linear-time algorithm is $\mathcal{O}(n^{-\alpha \log n})$, for some $\alpha > 0$, almost surely, when the processing times are in $[0, 1]$ and the density function is reasonably smooth. Tsai [Tsa95] modified the algorithm of Karmarkar and Karp in order to generate cardinality-balanced schedules preserving the asymptotic result on the average-case performance.

Beside many contributions concerning the average-case performance of several differencing strategies, results on the worst-case performance are rather seldom. Fischetti and Martello [FM87] showed that LDM has a performance ratio of $\frac{7}{6}$ for the special case that $m = 2$. For $m \geq 3$, Michiels [Mic04] proved in his Ph.D. project that the performance ratio of LDM concerning makespan-minimization is bounded between $\left(\frac{4}{3} - \frac{1}{3(m-1)}\right)$ and $\left(\frac{4}{3} - \frac{1}{3m}\right)$. Considered from a worst-case perspective, this means that LDM is worse than Multifit, but at least as good as LPT.

2.3 Bounds on the Maximum and Minimum Completion Time of LS-Schedules

In Section 2.1.2, we mentioned the performance ratio of the LS-algorithm compared to optimal scheduling concerning C_{max}-minimization as well as C_{min}-maximization. In this section, we want to contribute to this issue by deriving an upper bound on the value of C_{max} and a lower bound on the value of C_{min} for a given LS-heuristic. The main difference to Section 2.1.2 is that we try to express the bounds as a sum of processing times.

Therefore, we need some more notation. In Section 1.3 on page 6 it is mentioned that the superscript * refers to optimal scheduling. In addition to this, if we consider a certain algorithm A (A stands for an approximation algorithm or a heuristic) we use the superscript ^{A}, e. g., C_{max}^{LPT} denotes the makespan of LPT-scheduling.

In this context, we also generalize the term completion time of a machine, introduced on page 4, in order to compare completion times after the assignment of a subset of $\mathcal{J}$, i. e., after a partial schedule is constructed by an algorithm A. With $C_i^A(j)$ we denote the completion time of machine i after the assignment of the first j jobs according to algorithm A. Equivalently, we use the terms **current completion time** and **completion time so far** for it. With $C_{[i]}^A(j)$ we denote the current completion of the i-th shortest running machine after the assignment of the first j jobs according to A. Hence, $C_{[1]}^A(j) = C_{min}^A(j)$, $C_{[m]}^A(j) = C_{max}^A(j)$ and $C_{[m]}^A(j) - C_{[1]}^A(j) = C_\Delta^A(j)$. As an example, $C_{max}^{LPT}(j)$ is the current makespan after the assignment of the first j jobs according to the LPT-rule. If it is clear from the context, we omit the word *current*. Instead of $C_{max}^A(n)$, $C_{min}^A(n)$ or $C_\Delta^A(n)$ we often write C_{max}^A, C_{min}^A or C_Δ^A for short.

Lemma 2.3.1
Let L be an arbitrary priority list of the LS-algorithm and assume $1 \leq k \leq n - m$. Then,

(i) $C_{max}^L(k+m) \leq C_{max}^L(k) + \max\left\{t_{L(k+1)}, \ldots, t_{L(k+m)}\right\}.$

(ii) $C_{min}^L(k+m) \geq C_{min}^L(k) + \min\left\{t_{L(k+1)}, \ldots, t_{L(k+m)}\right\}.$

Proof

The two statements of the theorem are quite obvious. So, we omit the proof here.   ∎

Application of Lemma 2.3.1 leads to the following upper bound on C_{max}^L and the lower bound on C_{min}^L.

Corollary 2.3.2
Let L be an arbitrary priority list of the LS-algorithm. Then,

(i) $C_{max}^L(n) \leq \sum_{k=0}^{\frac{n}{m}-1} \max\left\{t_{L(km+1)}, \ldots, t_{L(km+m)}\right\}.$

(ii) $C_{min}^L(n) \geq \sum_{k=0}^{\frac{n}{m}-1} \min\left\{t_{L(km+1)}, \ldots, t_{L(km+m)}\right\}.$

2.4 Comparison of the SPT-LS- and LPT-Rule

The comparison of the two LS-heuristics SPT and LPT starts with a look at the structure of the schedules generated by SPT-LS. Then, we use this result to compare the objective function values of these two LS-heuristics afterwards.

Since SPT-scheduling is often defined more generally in literature, in this section we explicitly use SPT-LS whenever we mean the SPT-LS-rule in order to avoid confusion. The following definition of SPT-scheduling is quite common in literature [BCS74], [CMM67], [CS76]: Assign the jobs rank by rank, in order of decreasing rank, each job of a rank being assigned to a different machine. In concrete, jobs in rank $\frac{n}{m}$ being assigned first, jobs in rank $\frac{n}{m} - 1$ being assigned second, etc., jobs in rank 1 being assigned last. Note that the schedule generated by SPT-scheduling is cardinality-balanced. Since there are $m!$ ways of assigning the jobs in any given rank, $m!^{\frac{n}{m}}$ schedules are obtainable according to SPT-scheduling.

It is known [CS76] that widely different makespans may occur among SPT-schedules. Furthermore, it is well known [CMM67] that each SPT-schedule minimizes mean flow (or finishing) time. The mean flow time of a schedule is the sum of the finishing times of all n jobs (divided by n) and provides a measure of the average time that a job spends within the system. It also tends to reflect the number of unfinished jobs in the system.

The analysis made in the next subsection reveals that the SPT-LS-schedule fulfills the properties of SPT-schedules (as defined above) when a certain tie-break rule is applied. For this reason, SPT-LS is interesting to be studied.

2.4.1 Structure of SPT-LS-Schedules

Theorem 2.4.1
Whenever a job is assigned to a machine according to SPT-LS, then this machine has maximum completion time so far afterwards.

Proof
Assume that $1 \leq k < n$ jobs have already been assigned according to SPT-LS, and let the next job on the priority list of SPT-LS be assigned to machine i_{min}. This means the following:

- $C_{i_{min}}(k) = \min_{i=1,\dots,m} C_i(k)$

- $C_{i_{min}}(k+1) = C_{i_{min}}(k) + t_{L(k+1)}$

- $C_i(k+1) = C_i(k)$ for $i = 1, \dots, m$ and $i \neq i_{min}$.

Now, consider a machine with maximum completion time after the assignment of k jobs, and consider the last job that is assigned to that machine so far. Let this be machine i_{max} and job $L(\bar{k})$ $(1 \leq \bar{k} \leq k)$. Then, we know:

- $C_{i_{max}}(k) = \max_{i=1,\dots,m} C_i(k)$

- $C_{i_{max}}(k) = C_{i_{max}}(\bar{k})$

- $C_{i_{max}}(\bar{k} - 1) = \min_{i=1,\dots,m} C_i(\bar{k} - 1)$

- $C_{i_{max}}(\bar{k}) = C_{i_{max}}(\bar{k} - 1) + t_{L(\bar{k})}$

- $C_i(\bar{k}) = C_i(\bar{k} - 1)$ for $i = 1, \dots, m$ and $i \neq i_{max}$.

By the first two facts we derive the inequality chain $C_{i_{max}}(\bar{k}) = C_{i_{max}}(k) \geq C_i(k) \geq C_i(\bar{k})$ for all machines i. By the last three facts it follows that $C_{i_{max}}(\bar{k}) - C_i(\bar{k}) \leq t_{L(\bar{k})}$ for all i. Furthermore, by fact two and $C_i(\bar{k}) \leq C_i(k)$

it follows that $C_{i_{max}}(k) - C_i(k) \le t_{L(\bar{k})}$ for all i. Using the last inequality for $i = i_{min}$ yields the desired result:

$$C_{i_{min}}(k+1) = C_{i_{min}}(k) + t_{L(k+1)} \ge C_{i_{max}}(k) - t_{L(\bar{k})} + t_{L(k+1)}$$
$$= C_{i_{max}}(k) + \underbrace{t_{L(k+1)} - t_{L(\bar{k})}}_{\ge 0} \ge C_{i_{max}}(k).$$

■

Note that if the processing times are pairwise different, the last "$\ge$"-sign becomes a "$>$"-sign. This means that whenever a job is assigned to a machine according to SPT-LS, the next $(m-1)$ jobs are not assigned to this machine. Hence, the following machine completion times occur in the SPT-LS-schedule:

- the completion time of the makespan-machine is
  $$C_{max}^{SPT-LS} = t_1 + t_{m+1} + \ldots + t_{n-m+1} = \sum_{j=0}^{\frac{n}{m}-1} t_{jm+1}$$

- the completion time of the second longest running machine is
  $$C_{[m-1]}^{SPT-LS} = t_2 + t_{m+2} + \ldots + t_{n-m+2} = \sum_{j=0}^{\frac{n}{m}-1} t_{jm+2}$$

  $\vdots$

- the completion time of the shortest running machine is
  $$C_{min}^{SPT-LS} = t_m + t_{2m} + \ldots + t_n = \sum_{j=0}^{\frac{n}{m}-1} t_{jm+m}$$

In short, the completion time of the i-th longest running machine is

$$C_{[m-i+1]}^{SPT-LS} = t_i + t_{m+i} + \ldots + t_{n-m+i} = \sum_{j=0}^{\frac{n}{m}-1} t_{jm+i}$$

for $i = 1, \ldots, m$. In order to keep to non-permuted schedules, machine i runs i-th longest in the SPT-LS-schedule. In case of pairwise different processing times we have $C_1^{SPT-LS} > C_2^{SPT-LS} > \ldots > C_m^{SPT-LS}$.

This can be transfered to the more general case where two or more jobs have equal processing time. As mentioned earlier, the application of a certain tie-break rule in the LS-algorithm does not influence the values of $C_{max}, C_{min}, C_\Delta$ or the mean flow time but it has influence on the set of jobs that are processed by the same machine. Throughout this thesis it is our convention to apply the following rule in tie-situations at SPT-LS-scheduling: Assign the next job on the priority list to the machine that does not process any of the last $(m-1)$ jobs that have been assigned. In other words, we choose the machine whose current last job has the largest index of all current last jobs. Further, we claim that the first $(m-1)$ jobs of the priority list are assigned to different machines. With this rule, every m-th job of the list is processed by the same machine and no two jobs of the same rank are assigned to the same machine. The following corollary summarizes this subsection.

Corollary 2.4.2
In a (non-permuted) SPT-LS-schedule, the completion time of machine i is $C_i^{SPT-LS} = \sum_{j=0}^{\frac{n}{m}-1} t_{jm+i}$ for $i = 1, \ldots, m$.

2.4.2 Objective Function Values

Preliminaries

Studies on the makespan of SPT-schedules are contained for instance in [BCS74], [Cof73] and [CS76]. Usually, widely different makespans occur among the large number of SPT-schedules (note that the SPT-LS-schedule is only one special case of SPT-schedules). Let SPT* refer to a shortest SPT-schedule. Compared with the makespan of optimal scheduling, the following performance ratio is presented in [BCS74]:

$$\frac{C_{max}^{SPT^*}}{C_{max}^*} \leq 2 - \frac{1}{m}.$$

This bound is best possible as one can construct instances that approach this bound arbitrarily closely for any fixed m [BCS74]. However, in Section 2.6.3 we will show that no instance exists so that the makespan of the shortest SPT-schedule is exactly $(2 - \frac{1}{m})$ times the optimal makespan.

Regarding SPT-LS-scheduling as one special case of SPT-scheduling (applying the tie-break rule mentioned in the previous subsection), the $(2 - \frac{1}{m})$-bound is tight for any fixed m (cf. Theorem 2.1.1 on page 16). Moreover, we want to remark the following. In the SPT-LS-schedule, the i-th longest jobs of all ranks are processed by the same machine (and no other jobs are processed by this machine). Hence, the makespan of any SPT-schedule is not worse than the makespan of the SPT-LS-schedule and the same is true considering C_{min}, i. e., $C_{max}^{SPT} \leq C_{max}^{SPT-LS}$ and $C_{min}^{SPT} \geq C_{min}^{SPT-LS}$. Consequently, $C_\Delta^{SPT} \leq C_\Delta^{SPT-LS}$ holds.

Comparing the makespan of a shortest SPT-schedule with the makespan of the LPT-schedule, Bruno, Coffman, and Sethi [BCS74] as well as Coffman

[Cof73] showed the following bounds:

$$\frac{4m-1}{5m-2} \leq \frac{C_{max}^{SPT^*}}{C_{max}^{LPT}} \leq 2 - \frac{1}{m}.$$

The lower bound is tight for any fixed m as presented in [Cof73], and the upper bound is best possible as presented in [BCS74]. In Section 2.6.3 it will be revealed that the upper bound cannot be reached exactly. However, it is interesting to find that the same upper bound applies when LPT-scheduling is considered instead of optimal scheduling. The result concerning the lower bound is also interesting as the shortest SPT-schedule may be up to 20% shorter in the limit than the LPT-schedule.

The question whether job-systems exist such that the SPT-LS-schedule is shorter than the LPT-schedule, i. e., $C_{max}^{SPT-LS} < C_{max}^{LPT}$, is answered next.

Theoretical Results Comparing LPT and SPT-LS

To avoid triviality, we assume $\frac{n}{m} \geq 2$ throughout this thesis when we compare different LS-heuristics, unless anything else is mentioned. The case $m = n$ is simple: LS assigns the jobs with positive processing time to distinct machines. Hence, we get $C_{max}^{LS} = t_1, C_{min}^{LS} = t_m$ and $C_{\Delta}^{LS} = t_1 - t_m$ for each of the $m!$ different priority lists of the LS-algorithm.

Theorem 2.4.3
Consider an arbitrary instance of the IMSP. Then, the following three statements hold:

(i) *The SPT-LS-schedule is at least as long as the LPT-schedule, i. e., $C_{max}^{SPT-LS} \geq C_{max}^{LPT}$.*

(ii) *The C_{min}-value of the SPT-LS-schedule is at most as large as the C_{min}-value of the LPT-schedule, i. e., $C_{min}^{SPT-LS} \leq C_{min}^{LPT}$.*

(iii) *The C_{Δ}-value of the SPT-LS-schedule is at least as large as the C_{Δ}-value of the LPT-schedule, i. e., $C_{\Delta}^{SPT-LS} \geq C_{\Delta}^{LPT}$.*

Proof
To prove statement (i), apply Corollary 2.3.2 to the LPT-rule. Together with Corollary 2.4.2, this yields

$$C_{max}^{LPT} \leq t_1 + t_{m+1} + \ldots + t_{n-m+1} = C_{max}^{SPT-LS}.$$

The proof of statement (ii) works analogously. Again, application of Corollary 2.3.2 to LPT and Corollary 2.4.2 yield

$$C_{min}^{LPT} \geq t_m + t_{2m} + \ldots + t_n = C_{min}^{SPT-LS}.$$

Statement (iii) follows directly from the statements (i) and (ii). ∎

Summarizing the theoretical results on the (absolute) performance of SPT-LS-schedules compared to LPT-schedules we can state that SPT-LS is **dominated** by LPT in terms of C_{max}-minimization, C_{min}-maximization and C_Δ-minimization. It is easy to see that for any fixed number $m \geq 2$ of machines instances exist so that the LPT-schedule is better than the SPT-LS-schedule, in fact for all three objectives (cf. Theorem 2.1.1 on page 16). This justifies to speak of *dominance*.

On page 28, bounds on the ratio of the makespan of a shortest SPT-schedule and the makespan of an LPT-schedule are given. Considering the bounds for SPT-LS-scheduling instead of SPT-scheduling we get

$$1 \leq \frac{C_{max}^{SPT-LS}}{C_{max}^{LPT}} \leq 2 - \frac{1}{m}.$$

The bounds are tight for any fixed number m of machines. The upper bound is obtained by the same instance as used in the proof of Theorem 2.1.1. Here, it becomes clear that long jobs cause larger variations in the current completion times of the machines than short jobs. Hence, it is intuitively more promising to assign the long jobs before the short jobs like in LPT-scheduling and to use the short jobs to reduce variation in the completion times afterwards.

2.5   A Deeper Look at the SPT-LS-rule

This section consists of two main subsections. In Subsection 2.5.1, we study in detail the question whether priority rules exist that are dominated by the SPT-LS-rule. This will be done for all three objective functions.
Subsection 2.5.2 contains some results on the average-case performance of the SPT-LS-rule.

2.5.1　On the Dominance of the SPT-LS-rule

As mentioned in the previous section, the assignment of the jobs in non-decreasing order is one drawback of the SPT-LS-rule in terms of C_{max}-minimization as well as C_{min}-maximization. Hence, the following question arises quite naturally. Is the SPT-LS-rule the worst LS-heuristics for at least one of the three objectives studied in this thesis?

Before starting the investigation of this issue it is useful to introduce the term **group** or more precisely m-group (of jobs) which is to be distinguished from the term rank (introduced on page 3). Our understanding of m-groups in connection with priority lists is the following: m-groups simply lead to a partitioning of a priority list L, with jobs $L(gm + 1), \ldots, L(gm + m)$ in m-group $(g + 1)$ of L for $g = 0, 1, \ldots, \frac{n}{m} - 1$. So, the first m-group of a priority list L contains the first m jobs in L, the second m-group contains the next m jobs in L, etc. The last m-group contains the last m jobs in L. We want to remark that the two terms *ranks* and *m-groups* should not be mixed up. As jobs are *ranked* according to their processing times, they are *grouped* according to their position in a given priority list. Hence, jobs that belong to the same rank may appear in different m-groups.

Example 2.5.1
Consider the priority list $L = (1, 4, 7, 2, 5, 8, 3, 6, 9)$ and assume $m = 3$. Then, rank 1 contains the jobs J_1, J_2, J_3, rank 2 contains the jobs J_4, J_5, J_6, and rank 3 contains the jobs J_7, J_8, J_9. In contrast, m-group 1 contains the jobs J_1, J_4, J_7, m-group 2 contains the jobs J_2, J_5, J_8, and m-group 3 contains the jobs J_3, J_6, J_9.

Regarding the LPT- or the SPT-LS-rule for instance, jobs which belong to the same rank also build an m-group. Obviously, rank k and group k contain the same jobs for all $k = 1, \ldots, \frac{n}{m}$ considering the LPT-rule. Regarding the SPT-LS-rule, rank k and group $\left(\frac{n}{m} - k + 1\right)$ contain the same jobs for all k.

In connection with an m-group we introduce the term **group-minimum** by which the job with the largest index in the corresponding m-group is meant. So, this job has the shortest processing time in its group. On occasion, we also use the term group-minimum for the shortest processing time among the group-members. The term **group-maximum** is defined analogously. If confusion is unlikely we use the term group instead of m-group.

2.5.1.1 Analysis Concerning C_{max}

We begin with an analysis concerning the C_{max}-minimization problem.

Lemma 2.5.2
Let L be a priority list of the LS-algorithm with the following two properties:

- $C_{min}^L(n-m) \geq t_{m+1} + t_{2m+1} + \ldots + t_{n-m+1}$

- $L^{-1}(1) = n - m + k$ *for some* $k \in \{1, \ldots, m\}$.

Then, for each (m, n)-instance the inequality $C_{max}^{SPT-LS} \leq C_{max}^L$ holds.

Proof
The proof is simple. Assume $C_{min}^L(n-m) \geq t_{m+1} + t_{2m+1} + \ldots + t_{n-m+1}$ and assume the longest job to be placed at position $(n - m + k)$ in L for some $k \in \{1, \ldots, m\}$, i. e., $L(n - m + k) = 1$. Then, we can conclude:

$$C_{max}^L(n) \geq C_{min}^L(n - m + k - 1) + t_1 \geq C_{min}^L(n - m) + t_1 \geq$$
$$\geq t_1 + t_{m+1} + \ldots + t_{n-m+1} = C_{max}^{SPT-LS}(n).$$

■

Example 2.5.3
One prominent priority list that fulfills the two properties of Lemma 2.5.2 is $L = (2, 3, \ldots, n, 1)$. This rule is easy to remember since the order of the jobs is just like in LPT-scheduling except that the longest job is positioned last.

Corollary 2.5.4
There exist at least $(\frac{n}{m} - 1)! \, (m!)^{\frac{n}{m}}$ different priority lists L of the LS-algorithm so that the inequality $C_{max}^{SPT-LS} \leq C_{max}^L$ holds for each (m, n)-instance.

Proof
We show that at least $(\frac{n}{m} - 1)! \, (m!)^{\frac{n}{m}}$ different priority rules fulfill the two properties of Lemma 2.5.2. To prove this, we apply statement (ii) of Corollary 2.3.2 which ensures

$$C_{min}^L(n - m) \geq \sum_{k=0}^{\frac{n}{m}-2} \min\{t_{L(km+1)}, \ldots, t_{L(km+m)}\}$$

for any priority list L. This means that each priority list L whose set of group-minima

$$\Big\{ \min\{L(1), \ldots, L(m)\},$$
$$\min\{L(m+1), \ldots, L(2m)\},$$
$$\vdots$$
$$\min\{L(n-2m+1), \ldots, L(n-m)\} \Big\}$$

equals the set $\{m+1, 2m+1, \ldots, n-m+1\}$ fulfills the first property of Lemma 2.5.2.

The second property of Lemma 2.5.2 claims that the longest job of the job-system is positioned somewhere in the last m-group. Taking both into account, we can conclude the following concerning the structure of the m-groups of the priority lists that surely fulfill Lemma 2.5.2:

The last m-group must contain job J_1. We get the remaining $(m-1)$ jobs of this group after all other m-groups have been considered. So, in order to ensure that the set of group-minima equals the set $\{m+1, 2m+1, \ldots, n-m+1\}$, the first $(\frac{n}{m} - 1)$ m-groups (arbitrarily ordered) have to be:

$$\{J_2, J_3, \ldots, J_{m+1}\}, \{J_{m+2}, \ldots, J_{2m+1}\}, \ldots, \{J_{n-2m+2}, \ldots, J_{n-m+1}\}.$$

Hence, the jobs $J_{n-m+2}, \ldots, J_n$ are the remaining $(m-1)$ jobs which build the last m-group together with job J_1.

As mentioned before, all other groups can be ordered arbitrarily to fulfill the first property of Lemma 2.5.2. The jobs of each of these $(\frac{n}{m} - 1)$ m-groups can also be arbitrarily ordered as well as the jobs of the last m-group. This leads to a total number of at least $(\frac{n}{m} - 1)! \, (m!)^{\frac{n}{m}}$ different priority lists of the LS-algorithm that fulfill the two properties of Lemma 2.5.2.   ∎

Note that it is not clear if these are all priority lists L that fulfill the inequality $C_{max}^{SPT-LS} \leq C_{max}^{L}$ for each (m, n)-instance.

In a next step, we investigate the question whether SPT-LS is dominating each rule derived in the proof of the previous corollary. In other words, does there exist at least one (m, n)-instance for each of those rules and for each integer fraction $\frac{n}{m} \geq 2$ so that strict inequality $C_{max}^{SPT-LS} < C_{max}^{L}$ holds? This is answered in two parts.

The first part considers all priority lists L derived in the proof of Corollary 2.5.4 where J_1 is not the first job in the last m-group, i. e., $L^{-1}(1) > n - m + 1$. All of those rules are dominated by SPT-LS in terms of makespan-minimization as the following job-system shows:

$$t_1 = t_2 = \ldots = t_m = 3,$$
$$t_{m+1} = t_{m+2} = \ldots = t_n = 2.$$

For each rule under consideration we can state that after the assignment of $(n - m)$ jobs one machine has current completion time $2(\frac{n}{m} - 1)$ and all remaining $(m-1)$ machines have current completion time $2(\frac{n}{m}-1)+1$. Hence, job 1 is assigned to a machine which has completion time $2(\frac{n}{m} - 1) + 1$ so far. Thus, the makespan is $2(\frac{n}{m} + 1)$, whereas SPT-LS generates a schedule with makespan $2(\frac{n}{m} + 1) - 1$.

Example 2.5.5
Consider the $(3, 12)$-instance with processing times $t_1 = t_2 = t_3 = 3, t_4 = \ldots = t_{12} = 2$ and $L = (5, 7, 6, 4, 2, 3, 8, 9, 10, 12, 1, 11)$. Then, we get $C_{max}^L = 10$ and $C_{max}^{SPT-LS} = 9$.

The second part considers all remaining priority lists derived in the proof of Corollary 2.5.4. These are the lists where J_1 is the first job in the last m-group, i. e., $L^{-1}(1) = n - m + 1$. Here, the analysis is a bit more complicated and consists of three cases. We distinguish between different orders of the m-groups as well as different positions of the group-minima in the corresponding m-groups.

The following theorem deals with the *first* case.

Theorem 2.5.6
Let L be a priority list of the LS-algorithm with the following properties:

- *the group-minimum in the k-th m-group is J_{n-km+1} for $k = 1, \ldots, \frac{n}{m} - 1$*

- *the group-minimum of each of the m-groups $2, \ldots, \frac{n}{m} - 1$ appears at position 1 in its group, i. e., $L((k - 1)m + 1) = n - km + 1$ for $k = 2, \ldots, \frac{n}{m} - 1$*

- $L^{-1}(1) = n - m + 1.$

Then, equality $C_{max}^{SPT-LS} = C_{max}^L$ holds for each (m, n)-instance.

Hence, the first of the three cases is the one without dominance.

Proof

As t_{n-m+1} is positive and all m-group-minima of the groups $2, \ldots, \frac{n}{m} - 1$ are placed at position one in their group and the processing times are non-decreasing from group to group in L (group $\frac{n}{m}$ excluded), we can conclude that

$$C_{max}^L(km+1) = C_{min}^L(km) + t_{L(km+1)} > C_{max}^L(km)$$

for $k = 1, \ldots, \frac{n}{m} - 1$ and equality $C_{min}^L(km) = C_{max}^L((k-1)m+1)$ holds for $k = 1, \ldots, \frac{n}{m} - 1$. Hence, we have $C_{min}^L(km) = t_{n-m+1} + t_{n-2m+1} + \ldots + t_{n-km+1}$. By statement (i) of Corollary 2.3.2 we get a trivial upper bound on $C_{max}^L(n-m)$, namely $C_{max}^L(n-m) \leq t_{n-2m+2} + \ldots + t_{m+2} + t_2$. This leads us to an upper bound on $C_{\Delta}^L(n-m)$:

$$C_{\Delta}^L(n-m) \leq t_{n-2m+2} - t_{n-m+1} + t_{n-3m+2} - t_{n-2m+1} + - \ldots + t_2 - t_{m+1} =$$
$$= -t_{n-m+1} + (t_{n-2m+2} - t_{n-2m+1}) + \ldots + (t_{m+2} - t_{m+1}) + t_2 \leq$$
$$\leq t_2 - t_{n-m+1}.$$

So, after the assignment of $(n-m)$ jobs according to L the difference between the longest running machine so far and the shortest running machine so far is at most $t_2 - t_{n-m+1}$. As the next job to be assigned is job 1, we can conclude:

$$C_{max}^L(n-m+1) = C_{min}^L(n-m) + t_1 \quad \text{and}$$
$$C_{max}^L(n-m+1) - C_{max}^L(n-m) \geq t_1 - t_2 + t_{n-m+1} \geq t_{n-m+1}.$$

Furthermore, the processing time of each of the remaining $(m-1)$ jobs is at most t_{n-m+1}. Then, it follows

$$C_{max}^L(n) = C_{max}^L(n-m+1) = t_1 + t_{m+1} + \ldots + t_{n-m+1} = C_{max}^{SPT-LS}.$$

∎

As we have seen that each priority list L of Theorem 2.5.6 generates a schedule with the same makespan as the SPT-LS-schedule, this is not true concerning the minimum completion time (in case $m \geq 3$). We give a general instance on page 36.

The total number of priority lists that fulfill all properties of Theorem 2.5.6 is $m(m-1)!^{\frac{n}{m}}$.

Concerning the remaining two cases we assume $\frac{n}{m} \geq 3$. We will see that SPT-LS is dominating each of the corresponding priority lists in terms of makespan-minimization.

The *second* case considers priority lists L that are closely related to the first case. Again, we assume that $L^{-1}(1) = n - m + 1$, that the m-group-minima of the first $\left(\frac{n}{m} - 1\right)$ groups are the jobs $m + 1, 2m + 1, \ldots, n - m + 1$ and that the group-minima are placed at position one in the groups $2, \ldots, \frac{n}{m} - 1$. Furthermore, and this is the difference to the first case, we assume that there exists an m-group k $(k \in \{2, \ldots, \frac{n}{m} - 1\})$ so that $L((k-1)m+1) \neq n - km + 1$. In other words, we consider the same m-groups as in the first case and allow any order of the first $\left(\frac{n}{m} - 1\right)$ groups except the one order that is considered in case one.

Concerning the second case assume that $g \in \{2, \ldots, \frac{n}{m} - 1\}$ is the first m-group so that $L((g-2)m+1) < L((g-1)m+1)$. Note that such a group must exist in L, and consider the following job-system:

$$t_1 = \ldots = t_{(\max\{L((g-2)m+1),\ldots,L((g-2)m+m)\}-1)} = 4,$$

$$t_{\max\{L((g-2)m+1),\ldots,L((g-2)m+m)\}} = \ldots = t_{(\min\{L((g-1)m+1),\ldots,L((g-1)m+m)\}-1)} = 2,$$

$$t_{\min\{L((g-1)m+1),\ldots,L((g-1)m+m)\}} = \ldots = t_n = 1.$$

We get $C^L_{min}(gm) > t_{\max\{L(1),\ldots,L(m)\}} + \ldots + t_{(g-1)m+1}$ which leads to $C^L_{min}(n-m) > t_{m+1} + \ldots + t_{n-m+1}$ (cf. Corollary 2.3.2 on page 24) and $C^L_{max}(n) > C^{SPT-LS}_{max}(n)$, consequently.

Example 2.5.7
Consider the $(3, 12)$-instance with processing times $t_1 = t_2 = t_3 = 4, t_4 = \ldots = t_7 = 2, t_8 = \ldots = t_{12} = 1$ and $L = (7, 6, 5, 4, 2, 3, 10, 8, 9, 1, 12, 11)$. Here, we have $g = 3$ since $L(4) = 4 < 10 = L(7)$. The makespan of the L-schedule is $C^L_{max} = 10$, whereas the makespan of the SPT-LS-schedule is $C^{SPT-LS}_{max} = 9$.

In the *third* and last case, we consider all priority lists L derived in Corollary 2.5.4 with $L^{-1}(1) = n - m + 1$ and at least one group-minimum of the m-groups $2, \ldots, \frac{n}{m} - 1$ being not placed at position one in its group.

Assume that $g \in \{2, \ldots \frac{n}{m} - 1\}$ is the first m-group in L in which the group-minimum is not placed at position one in group g. We distinguish two sub-cases for the construction of the job-system.

In sub-case one we assume the jobs in group g to have smaller indices than the jobs in group $(g - 1)$. For it, consider the following job-system:

$$t_1 = \ldots = t_{(\max\{L((g-1)m+1),\ldots,L((g-1)m+m)\}-1)} = 4,$$

$$t_{\max\{L((g-1)m+1),\ldots,L((g-1)m+m)\}} = \ldots = t_{(\max\{L((g-2)m+1),\ldots,L((g-2)m+m)\}-1)} = 3,$$

$$t_{\max\{L((g-2)m+1),\ldots,L((g-2)m+m)\}} = \ldots = t_n = 2.$$

Then, strict inequality

$$C_{min}^L(gm) > t_{\max\{L(1),...,L(m)\}} + t_{L(m+1)} + \ldots +$$
$$+ t_{L((g-2)m+1)} + t_{\max\{L((g-1)m+1),...,L((g-1)m+m)\}}$$

holds which leads to $C_{min}^L(n-m) > t_{m+1} + \ldots + t_{n-m+1}$.

Example 2.5.8
Consider the $(3,12)$-instance with processing times $t_1 = t_2 = t_3 = 4, t_4 = t_5 = t_6 = 3, t_7 = \ldots = t_{12} = 2$ and $L = (5,7,6,3,2,4,10,8,9,1,11,12)$. Here, we have $g = 2$ since $L(4) = 3 < 4 = L(6)$. The makespan of the L-schedule is $C_{max}^L = 12$, whereas the makespan produced by SPT-LS is $C_{max}^{SPT-LS} = 11$.

In sub-case two we assume the jobs in group g to have larger indices than the jobs in group $(g-1)$. For it, consider the following job-system:

$$t_1 = \ldots = t_{(\max\{L((g-2)m+1),...,L((g-2)m+m)\}-1)} = 4,$$
$$t_{\max\{L((g-2)m+1),...,L((g-2)m+m)\}} = \ldots = t_{(\max\{L((g-1)m+1),...,L((g-1)m+m)\}-1)} = 3,$$
$$t_{\max\{L((g-1)m+1),...,L((g-1)m+m)\}} = \ldots = t_n = 2.$$

Then, strict inequality

$$C_{min}^L(gm) > t_{\max\{L(1),...,L(m)\}} + t_{L(m+1)} + \ldots +$$
$$+ t_{L((g-2)m+1)} + t_{\max\{L((g-1)m+1),...,L((g-1)m+m)\}}$$

holds which leads again to $C_{min}^L(n-m) > t_{m+1} + \ldots + t_{n-m+1}$.

Example 2.5.9
Consider the $(3,12)$-instance with processing times $t_1 = \ldots = t_6 = 4, t_7 = t_8 = t_9 = 3, t_{10} = t_{11} = t_{12} = 2$ and $L = (7,6,5,8,10,9,3,2,4,1,11,12)$. Here, we have $g = 2$ since $L(4) = 8 < 10 = L(5)$. The makespan produced by L is $C_{max}^L = 14$, whereas the makespan produced by SPT-LS is $C_{max}^{SPT-LS} = 13$.

We take a final look at the priority lists derived in Corollary 2.5.4 by comparing their C_{min}-value to C_{min}^{SPT-LS}. We assume $m \geq 3$ and will see that SPT-LS does not dominate any of those rules in terms of C_{min}-maximization. Therefore, consider the following (m,n)-instance:

$$t_1 = \ldots = t_{n-m-1} = 8,$$
$$t_{n-m} = \ldots = t_{n-2} = 4,$$
$$t_{n-1} = 2,$$
$$t_n = 1.$$

Then, after the assignment of $(n - m)$ jobs according to any L of Corollary 2.5.4 there are two machines with current completion time $8(\frac{n}{m} - 2) + 4$ and each of the remaining $(m-2)$ machines has current completion time $8(\frac{n}{m} - 1)$. From this, we can conclude the following inequality for any priority list L under consideration:

$$C^L_{min}(n) \geq 8(\frac{n}{m} - 2) + 4 + 2 > 8(\frac{n}{m} - 2) + 4 + 1 = C^{SPT-LS}_{min}(n).$$

2.5.1.2 Analysis Concerning C_{min}

In the next part we consider the main issue of Subsection 2.5.1 concerning the C_{min}-maximization problem. This is done analogously to the previous C_{max}-part.

Theorem 2.5.10
Let L be a priority list of the LS-algorithm with the following two properties:

- $C^L_{max}(n - m) \leq t_m + t_{2m} + \ldots + t_{n-m}$
- $L^{-1}(n) = n - m + k$ *for some* $k \in \{1, \ldots, m\}$.

Then, for each (m, n)-instance inequality $C^{SPT-LS}_{min} \geq C^L_{min}$ holds.

Proof
Assume $C^L_{max}(n - m) \leq t_m + t_{2m} + \ldots + t_{n-m}$ and assume the shortest job to be placed at position $(n - m + k)$ for some $k \in \{1, \ldots, m\}$ in L, i. e., $L^{-1}(n) = n - m + k$. We distinguish two cases.

Case 1: Each of the last m jobs of L is processed by a different machine.

Since job n is contained in the last m-group of L and each machine processes exactly one of these jobs, job n is assigned to the k-th shortest running machine after $(n - m)$ jobs assigned. Hence, after the assignment of all n jobs according to L there exists a machine with completion time $C^L_{[k]}(n - m) + t_{L(n-m+k)}$, and we can conclude:

$$C^L_{min}(n) \leq C^L_{[k]}(n - m) + t_{L(n-m+k)} = C^L_{[k]}(n - m) + t_n \leq$$
$$\leq C^L_{max}(n - m) + t_n \leq t_m + t_{2m} + \ldots + t_{n-m} + t_n = C^{SPT-LS}_{min}(n).$$

Case 2: At least one machine processes more than one of the last m jobs of L.

This means that after the assignment of all n jobs according to L there exists a machine with completion time $C_{max}^L(n-m)$. We can conclude:

$$C_{min}^L(n) \leq C_{max}^L(n-m) \leq C_{max}^L(n-m) + t_n \leq$$
$$\leq t_m + t_{2m} + \ldots + t_{n-m} + t_n = C_{min}^{SPT \ LS}(n).$$

■

Example 2.5.11
One priority list that fulfills the two properties of Theorem 2.5.10 is $L = (m, m+1, \ldots, n, 1, 2, \ldots, m-1)$. The order of the jobs is just like in LPT-scheduling except that the longest $(m-1)$ jobs are positioned at the end of the list.

Corollary 2.5.12
There exist at least $(\frac{n}{m} - 1)! \, (m!)^{\frac{n}{m}}$ different priority lists L of the LS-algorithm so that the inequality $C_{min}^{SPT-LS} \geq C_{min}^L$ holds for each (m, n)-instance.

Proof
We show that at least $(\frac{n}{m} - 1)! \, (m!)^{\frac{n}{m}}$ different priority rules fulfill the two properties of Theorem 2.5.10. The proof of this part works mainly analogous to the proof of Corollary 2.5.4 and is shortened a bit, therefore.

Applying statement (i) of Corollary 2.3.2 leads to

$$C_{max}^L(n-m) \leq \sum_{k=0}^{\frac{n}{m}-2} \max\{t_{L(km+1)}, \ldots, t_{L(km+m)}\}$$

for any priority list L. This means that each priority list L whose set of group-maxima

$$\Big\{ \max\{L(1), \ldots, L(m)\},$$
$$\max\{L(m+1), \ldots, L(2m)\},$$
$$\vdots$$
$$\max\{L(n-2m+1), \ldots, L(n-m)\}\Big\}$$

equals the set $\{m, 2m, \ldots, n-m\}$ fulfills the first property of Theorem 2.5.10. Taking also into account that the shortest job of the job-system is positioned

in the last m-group, the following priority lists fulfill Theorem 2.5.10:
The first $\left(\frac{n}{m} - 1\right)$ m-groups are

$$\{J_m, J_{m+1}, \ldots, J_{2m-1}\}, \{J_{2m}, \ldots, J_{3m-1}\}, \ldots, \{J_{n-m}, \ldots, J_{n-1}\}.$$

These groups can be ordered arbitrarily as well as the jobs of each of these groups. The last m-group contains the jobs $J_1, J_2, \ldots, J_{m-1}, J_n$ which can be ordered arbitrarily, too. This leads to a total number of at least $\left(\frac{n}{m}-1\right)! \, (m!)^{\frac{n}{m}}$ different priority lists of the LS-algorithm that fulfill the two properties of Theorem 2.5.10.   ∎

Again, it is not clear if these are all priority lists L that fulfill the inequality $C_{min}^{SPT-LS} \geq C_{max}^{L}$ for each (m, n)-instance.
Furthermore, note that the priority lists derived in the Corollaries 2.5.4 and 2.5.12 are identical in case $m = 2$. In the more general case $m \geq 3$, none of the priority lists derived in Corollary 2.5.4 appears in Corollary 2.5.12 and vice versa.

Similar to our analysis concerning C_{max}, we will investigate the question whether SPT-LS is dominating each rule derived in the proof of the previous Corollary 2.5.12. In other words, does there exist at least one (m, n)-instance for each of those rules and for each integer fraction $\frac{n}{m} \geq 2$ so that strict inequality $C_{min}^{SPT-LS} > C_{min}^{L}$ holds? This is answered in two parts.

The first part considers all priority lists L derived in the proof of Corollary 2.5.12 where J_n is not the last job in the last m-group, i. e., $L^{-1}(n) < n$. All of those rules are dominated by SPT-LS in terms of C_{min}-maximization as the following (m, n)-instance shows:

$$t_1 = t_2 = \ldots = t_{n-m} = 3,$$
$$t_{n-m+1} = \ldots = t_n = 2.$$

For each rule under consideration we can state that after the assignment of $(n - m)$ jobs one machine has current completion time $3(\frac{n}{m} - 1)$ and all remaining $(m-1)$ machines have current completion time $3(\frac{n}{m}-1)-1$. Hence, job n is assigned to a machine which has completion time $3(\frac{n}{m} - 1) - 1$ so far. Thus, the minimum completion time is $3(\frac{n}{m} - 1) + 1$, whereas SPT-LS generates a schedule with minimum completion time $3(\frac{n}{m} - 1) + 2$.

Example 2.5.13
Consider the $(3, 12)$-instance with processing times $t_1 = t_2 = \ldots = t_9 = 3, t_{10} = t_{11} = t_{12} = 2$ and $L = (6, 8, 7, 3, 4, 5, 11, 10, 9, 2, 12, 1)$. Then, we get $C_{min}^{L} = 10$ and $C_{min}^{SPT-LS} = 11$.

The second part considers all remaining priority lists derived in the proof of Corollary 2.5.12. These are the lists where J_n is the last job in the last m-group, i. e., $L^{-1}(n) = n$. For it, we distinguish three cases quite analogous to the corresponding investigations concerning the objective function C_{max}.

The next theorem deals with the *first* case.

Theorem 2.5.14
Let L be a priority list of the LS-algorithm with the following properties:

- *the group-maximum in the k-th m-group is J_{n-km} for $k = 1, \ldots, \frac{n}{m} - 1$*

- *the group-maximum of each of the m-groups $2, \ldots, \frac{n}{m} - 1$ appears at position m in its group, i. e., $L(km) = n - km$ for $k = 2, \ldots, \frac{n}{m} - 1$*

- *$L(n) = n$.*

Then, equality $C_{min}^{SPT-LS} = C_{min}^{L}$ holds for each (m, n)-instance.

Hence, the first of the three cases is the one without dominance.

Proof
The proof works analogously to the proof of Theorem 2.5.6. Since all m-group-maxima of the groups $2, \ldots, \frac{n}{m} - 1$ are placed at position m in their group and the processing times are non-decreasing from group to group (group $\frac{n}{m}$ excluded), we can conclude that

$$C_{max}^{L}(km) = C_{min}^{L}(km - 1) + t_{L(km)} = C_{max}^{L}((k-1)m) + t_{L(km)} =$$
$$= t_{n-m} + t_{n-2m} + \ldots + t_{n-km}$$

for $k = 1, \ldots, \frac{n}{m} - 1$.
By statement (ii) of Corollary 2.3.2 we get a trivial lower bound on $C_{min}^{L}(n - m)$, namely $C_{min}^{L}(n-m) \geq t_{n-1} + \ldots + t_{3m-1} + t_{2m-1}$. This leads to an upper bound on $C_{\Delta}^{L}(n - m)$:

$$C_{\Delta}^{L}(n - m) \leq t_m + (t_{2m} - t_{2m-1}) + \ldots + (t_{n-m} - t_{n-m-1}) - t_{n-1} \leq$$
$$\leq t_m - t_{n-1}.$$

So, after the assignment of $(n-m)$ jobs according to L the difference between the longest running machine so far and the shortest running machine so far is at most $t_m - t_{n-1}$. As each of the next $(m - 1)$ jobs to be assigned has a length of at least t_m, we can conclude that each of those jobs is assigned to a separate machine.

Since $t_j - t_m + t_{n-1} \geq t_{n-1}$ for $j \in \{1, \ldots, m-1\}$, the machine that processes job n is shortest running, i. e.,

$$C_{min}^L(n) = C_{min}^L(n-1) + t_{L(n)} = C_{max}^L(n-m) + t_n =$$
$$= t_m + t_{2m} + \ldots + t_n = C_{min}^{SPT-LS}(n).$$

∎

While we have seen that each priority list L of Theorem 2.5.14 generates a schedule with the same C_{min}-value as the SPT-LS-schedule, this is not true concerning the makespan of such schedules in case $m \geq 3$. We give a general instance on page 43.

The total number of priority lists that fulfill all properties of Theorem 2.5.14 is $m(m-1)!^{\frac{n}{m}}$.

Concerning the remaining two cases we assume $\frac{n}{m} \geq 3$. We will see that SPT-LS is dominating each of the corresponding priority lists in terms of C_{min}-maximization.

The *second* case considers priority lists L that are closely related to the first case. Again, we assume that $L(n) = n$, that the m-group-maxima of the first $(\frac{n}{m} - 1)$ groups are the jobs $m, 2m, \ldots, n-m$, and that the group-maxima of the groups $2, \ldots, \frac{n}{m} - 1$ are placed at position m. Furthermore, and this is the difference to the first case, we assume that there exists an m-group k ($k \in \{2, \ldots, \frac{n}{m} - 1\}$) so that $L(km) \neq n - km$.

Regarding the second case, assume that $g \in \{2, \ldots, \frac{n}{m} - 1\}$ is the first m-group so that $L(gm - m) < L(gm)$. Note that such a group must exist in L, and consider the following job-system:

$$t_1 = \ldots = t_{\min\{L((g-2)m+1),\ldots,L((g-2)m+m)\}} = 4,$$
$$t_{(\min\{L((g-2)m+1),\ldots,L((g-2)m+m)\}+1)} = \ldots = t_{\min\{L((g-1)m+1),\ldots,L((g-1)m+m)\}} = 2,$$
$$t_{(\min\{L((g-1)m+1),\ldots,L((g-1)m+m)\}+1)} = \ldots = t_n = 1.$$

We get $C_{max}^L(gm) < t_{\min\{L(1),\ldots,L(m)\}} + \ldots + t_{gm}$ which leads to $C_{max}^L(n-m) < t_m + \ldots + t_{n-m}$ (cf. Corollary 2.3.2 on page 24) and $C_{min}^L(n) < C_{min}^{SPT-LS}(n)$, consequently.

Example 2.5.15

Consider the $(3, 12)$-instance with processing times $t_1 = t_2 = t_3 = 4, t_4 = \ldots = t_9 = 2, t_{10} = t_{11} = t_{12} = 1$ and $L = (8, 6, 7, 5, 4, 3, 10, 11, 9, 2, 1, 12)$. Here, we have $g = 3$ since $L(6) = 3 < 9 = L(9)$. The minimum completion time of the L-schedule is $C_{min}^L = 8$, whereas the minimum completion time of the SPT-LS-schedule is $C_{min}^{SPT-LS} = 9$.

In the *third* and last case, we consider all priority lists L derived in Corollary 2.5.12 with $L(n) = n$ and at least one group-maximum of the m-groups $2, \ldots, \frac{n}{m} - 1$ being not placed at position m in its group.

Assume that $g \in \{2, \ldots \frac{n}{m} - 1\}$ is the first m-group in L in which the group-maximum is not placed at position m in group g. We distinguish two sub-cases for the construction of the job-system.

In sub-case one we assume the jobs in group g to have smaller indices than the jobs in group $(g-1)$. For it, consider the following job-system:

$$t_1 = \ldots = t_{\min\{L((g-1)m+1), \ldots, L((g-1)m+m)\}} = 4,$$
$$t_{(\min\{L((g-1)m+1), \ldots, L((g-1)m+m)\}+1)} = \ldots = t_{\min\{L((g-2)m+1), \ldots, L((g-2)m+m)\}} = 3,$$
$$t_{(\min\{L((g-2)m+1), \ldots, L((g-2)m+m)\}+1)} = \ldots = t_n = 2.$$

Then, strict inequality

$$C_{max}^L(gm) < t_{\min\{L(1), \ldots, L(m)\}} + t_{L(2m)} + \ldots +$$
$$+ t_{L(gm-m)} + t_{\min\{L((g-1)m+1), \ldots, L((g-1)m+m)\}}$$

holds which leads to $C_{max}^L(n-m) < t_m + \ldots + t_{n-m}$.

Example 2.5.16
Consider the $(3, 12)$-instance with processing times $t_1 = t_2 = t_3 = 4, t_4 = t_5 = t_6 = 3, t_7 = \ldots = t_{12} = 2$ and $L = (8, 7, 6, 4, 3, 5, 10, 11, 9, 2, 1, 12)$. Here, we have $g = 2$ since $L(6) = 5 > 3 = L(5)$. The minimum completion time of the L-schedule is $C_{min}^L = 10$, whereas the minimum completion time of the SPT-LS-schedule is $C_{min}^{SPT-LS} = 11$.

In sub-case two we assume the jobs in group g to have larger indices than the jobs in group $(g-1)$. For it, consider the following job-system:

$$t_1 = \ldots = t_{\min\{L((g-2)m+1), \ldots, L((g-2)m+m)\}} = 4,$$
$$t_{(\min\{L((g-2)m+1), \ldots, L((g-2)m+m)\}+1)} = \ldots = t_{\min\{L((g-1)m+1), \ldots, L((g-1)m+m)\}} = 3,$$
$$t_{(\min\{L((g-1)m+1), \ldots, L((g-1)m+m)\}+1)} = \ldots = t_n = 2.$$

Then, strict inequality

$$C_{max}^L(gm) < t_{\min\{L(1), \ldots, L(m)\}} + t_{L(2m)} + \ldots +$$
$$+ t_{L(gm-m)} + t_{\min\{L((g-1)m+1), \ldots, L((g-1)m+m)\}}$$

holds which leads again to $C_{max}^L(n-m) < t_m + \ldots + t_{n-m}$.

Example 2.5.17

Consider the $(3,12)$-instance with processing times $t_1 = \ldots = t_6 = 4, t_7 = t_8 = t_9 = 3, t_{10} = t_{11} = t_{12} = 2$ and $L = (8,7,6,10,9,11,5,4,3,1,2,12)$. Here, we have $g = 2$ since $L(6) = 11 > 9 = L(5)$. The minimum completion time of the L-schedule is $C_{min}^L = 12$, whereas the minimum completion time of the SPT-LS-schedule is $C_{min}^{SPT-LS} = 13$.

We take a final look at the priority lists L derived in Corollary 2.5.12 by comparing their makespan to the makespan of the SPT-LS-schedule. Here, we assume $m \geq 3$ and we will see that some of them generate schedules with the same makespan as the SPT-LS-schedule for any (m,n)-instance. Before presenting such LS-rules, we give a job-system for which most of the priority lists L derived in Corollary 2.5.12 (except the lists[4] that are explained next) yield a better makespan than SPT-LS. For it, consider the following job-system and assume each L under consideration fulfills either $L^{-1}(1) \notin \{n-1, n\}$ or $L^{-1}(n) < L^{-1}(1) = n-1$:

$$t_1 = 6,$$
$$t_2 = \ldots = t_{m+1} = 4,$$
$$t_{m+2} = \ldots = t_{n-1} = 2,$$
$$t_n = 1.$$

Then, after the assignment of $(n-m)$ jobs according to any L derived in Corollary 2.5.12 there are two machines with current completion time $\frac{2n}{m}$ and each of the remaining $(m-2)$ machines has current completion time $(\frac{2n}{m} - 2)$. Taking the assumptions on the position of job 1 in L into account, the following result can easily be verified:

$$C_{max}^L(n) \leq \frac{2n}{m} + 5 < \frac{2n}{m} + 6 = C_{max}^{SPT-LS}(n).$$

2.5.1.3 Analysis Concerning C_Δ

To close this subsection, we take a look at the objective function C_Δ. We assume $m \geq 3$. As indicated before, we are able to present priority lists L so that inequality $C_\Delta^{SPT-LS} \leq C_\Delta^L$ holds for each (m,n)-instance. Therefore,

---

[4]We take a detailed look at such priority lists in Section 2.5.1.3. They are candidates for being dominated by SPT-LS in terms of C_Δ-minimization.

we consider the same m-groups as derived in Corollary 2.5.12, namely

$$\{m, m+1, \ldots, 2m-1\},$$
$$\{2m, \ldots, 3m-1\},$$
$$\vdots$$
$$\{n-m, \ldots, n-1\},$$
$$\{n, 1, 2, \ldots, m-1\}.$$

Theorem 2.5.18
Let L be a priority list of the LS-algorithm with the following properties:

- *the group-maximum in the k-th m-group is J_{n-km} for $k = 1, \ldots, \frac{n}{m} - 1$*

- *the group-maximum of each of the m-groups $2, \ldots, \frac{n}{m} - 1$ appears at position m in its group, i. e., $L(km) = n - km$ for $k = 2, \ldots, \frac{n}{m} - 1$*

- *the job with the second smallest index in each of the m-groups $2, \ldots, \frac{n}{m} - 1$ appears at position $(m-1)$ in its group, i. e., $L(km-1) = n - km + 1$ for $k = 2, \ldots, \frac{n}{m} - 1$*

- *$L(n-1) = 1$ and $L(n) = n$.*

Then, equality $C_\Delta^{SPT-LS} = C_\Delta^L$ holds for each (m,n)-instance.

Proof
In each of the groups $2, \ldots, \frac{n}{m} - 1$ the job with smallest index is placed at position m and the job with second smallest index is placed at position $(m-1)$. Moreover, the processing times are non-decreasing from group to group (group $\frac{n}{m}$ excluded). Hence, the following current machine completion times are easily verified:

$$C_{[m]}^L(n-m) = t_m + t_{2m} + \ldots + t_{n-m} = C_{max}^L(n-m),$$
$$C_{[m-1]}^L(n-m) = t_{m+1} + t_{2m+1} + \ldots + t_{n-m+1}.$$

By Theorem 2.5.14 we know $C_\Delta^L(n-m) \leq t_m - t_{n-1}$. Thus, each of the next $(m-2)$ jobs is assigned to a separate machine, and job 1 is assigned to the machine with completion time $C_{[m-1]}^L(n-m)$ so far. This machine has the longest running time after the assignment of job 1. The assignment of job n is discussed in Theorem 2.5.14 on page 40.

So, we can conclude:

$$C_\Delta^L(n) = C_{[m-1]}^L(n-m) + t_1 - C_{[m]}^L(n-m) - t_n =$$
$$= t_1 + t_{m+1} + \ldots + t_{n-m+1} - t_m - t_{2m} - \ldots - t_n = C_\Delta^{SPT-LS}(n).$$

∎

The total number of priority lists L that fulfill the properties of Theorem 2.5.18 is $m(m-1)((m-2)!)^{\frac{n}{m}}$.

Theorem 2.5.19
Let L be a priority list of the LS-algorithm with the following properties:

- *the group-maximum in the k-th m-group is J_{n-km} for $k = 1, \ldots, \frac{n}{m} - 1$*

- *the group-maximum of each of the m-groups $2, \ldots, \frac{n}{m} - 1$ appears at position m in its group, i. e., $L(km) = n - km$ for $k = 2, \ldots, \frac{n}{m} - 1$*

- *the job with the second smallest index in each of the m-groups $2, \ldots, \frac{n}{m} - 1$ appears at position $(m-1)$ in its group, i. e., $L(km-1) = n - km + 1$ for $k = 2, \ldots, \frac{n}{m} - 1$*

- *$L(n-1) = n$ and $L(n) = 1$.*

Then, inequality $C_\Delta^{SPT-LS} \leq C_\Delta^L$ holds for each (m, n)-instance.

Proof
Since the first three properties are the same as the ones in Theorem 2.5.18 we know that job n is assigned to the machine with completion time $C_{[m-1]}^L(n-m) = t_{m+1} + t_{2m+1} + \ldots + t_{n-m+1}$. Depending on the difference $(C_{[m]}^L(n-m) - C_{[m-1]}^L(n-m) - t_n)$, we consider the following two cases.

Case 1: $C_{[m]}^L(n-m) - C_{[m-1]}^L(n-m) - t_n > 0$.
This means that job 1 is assigned to the same machine as job n is assigned to and we can conclude:

$$C_\Delta^L(n) = C_{[m-1]}^L(n-m) + t_n + t_1 - C_{[m]}^L(n-m) =$$
$$= t_1 + t_{m+1} + \ldots + t_{n-m+1} + t_n - t_m - t_{2m} - \ldots - t_{n-m} \geq$$
$$\geq C_\Delta^{SPT-LS}(n).$$

Case 2: $C^L_{[m]}(n-m) - C^L_{[m-1]}(n-m) - t_n \leq 0$.

This means that job 1 is assigned to the machine with completion time $C^L_{[m]}(n-m)$ so far and we can conclude:

$$C^L_{max}(n) = C^L_{[m]}(n-m) + t_1 = t_1 + t_m + t_{2m} + \ldots + t_{n-m} \geq$$
$$\geq C^{SPT-LS}_{max}(n).$$

Theorem 2.5.10 ensures $C^L_{min}(n) \leq C^{SPT-LS}_{min}(n)$ for any L under consideration. This leads to $C^{SPT-LS}_\Delta(n) \leq C^L_\Delta(n)$.

■

The total number of priority lists L that fulfill the properties of Theorem 2.5.19 is again $m(m-1)((m-2)!)^{\frac{n}{m}}$.

To verify that SPT-LS even dominates each priority list L that fulfills the properties of Theorem 2.5.19, consider the next job-system:

$$t_1 = \ldots = t_m = 3,$$
$$t_{m+1} = 2,$$
$$t_{m+2} = \ldots = t_n = 1.$$

It can easily be verified that any L which fulfills the properties of Theorem 2.5.19 generates a schedule with makespan $(4 + \frac{n}{m})$ and minimum completion time $(1 + \frac{n}{m})$. This yields $C^L_\Delta(n) = 3$, whereas SPT-LS generates a schedule with $C^{SPT-LS}_\Delta(n) = 1$.

2.5.2 Average-Case Analysis of the SPT-LS-rule

In Section 2.4.1 we dealt with the structure of SPT-LS-schedules. The results presented there provide a nice opportunity for an average-case analysis of the SPT-LS-rule. From Corollary 2.4.2 we know the completion time of the i-th longest running machine in the SPT-LS-schedule:

$$C^{SPT-LS}_{[m-i+1]} = \sum_{j=0}^{\frac{n}{m}-1} t_{jm+i} \quad \text{for} \quad i = 1, \ldots, m.$$

Before we present the results of our average-case analysis we introduce the term **order statistic**. Suppose that $X_1, \ldots, X_n$ are n independent, identically distributed (i. i. d.) random variables. The corresponding order statistics $X_{1:n} \geq X_{2:n} \geq \ldots \geq X_{n:n}$ are the X_j's arranged in non-increasing order.

$X_{j:n}$ is called the j-th largest order statistic.

Assuming that the processing times are independent samples from identically distributed random variables X_j, $X_{j:n}$ corresponds to the j-th longest processing time.

The result of Corollary 2.4.2 can be used to compute the expected value of the i-th longest running machine for all machines i, provided that the expected value of each order statistic is known. It is also important to note that in this section $n = qm + r$ ($q \in \mathbb{N}$) denotes the number of jobs of an instance.[5] Instead of $r \in \{0, 1, \ldots, m-1\}$, we allow $r \in \{1, \ldots, m\}$ for computational advantage.

The stochastic model considered here is that we assume the processing times to be independent samples, uniformly distributed in the unit interval $[0, 1]$. For it, the expected value of the corresponding j-th order statistic is known [ABN92] to be

$$\mathbb{E}[X_{j:n}] = \frac{n - j + 1}{n + 1}$$

for $j = 1, \ldots, n$.

Then, the expected completion time of the i-th longest running machine in the SPT-LS-schedule can be computed as follows:

$$\mathbb{E}[C_{[m-i+1]}^{SPT-LS}] = \begin{cases} \mathbb{E}[\sum_{k=0}^{q} X_{(km+i):n}] & \text{if } i \leq r, \\ \mathbb{E}[\sum_{k=0}^{q-1} X_{(km+i):n}] & \text{otherwise.} \end{cases}$$

In a next step we explicitly compute the expected makespan of the SPT-LS-schedule in the underlying stochastic model. Assuming any fixed number m of machines, we get

$$\mathbb{E}[C_{max}^{SPT-LS}] = \mathbb{E}[C_{[m]}^{SPT-LS}] = \mathbb{E}[\sum_{k=0}^{q} X_{(km+i):n}] = \sum_{k=0}^{q} \mathbb{E}[X_{(km+i):n}] =$$

$$= \sum_{k=0}^{q} \frac{n - (km+1) + 1}{n+1} = \frac{n(q+1)}{n+1} - \frac{mq(q+1)}{2(n+1)} =$$

$$= \frac{n + (n - qm) + qn + q(n - qm)}{2(n+1)} = \frac{(n+r)(q+1)}{2(n+1)} =$$

$$= \frac{(n+r)(n+m-r)}{2m(n+1)} \approx \frac{n}{2m} \quad (n \gg m, r).$$

---

[5]In our stochastic model, the probability is zero that a job has length zero. So, n denotes the number of non-dummy-jobs in our experiments.

In an analogous way, the expected values of C_{min}^{SPT-LS} and C_{Δ}^{SPT-LS} can be determined.

Next, we briefly compare the expected performance of the SPT-LS-rule to optimum scheduling. Therefore, we use the trivial lower bound on the optimum makespan $C_{max}^* \geq \frac{1}{m} \sum_{j=1}^{n} X_j$ given by Lemma 1.3.2 and again we assume m to be fixed. Then, a simple calculation of the expected (absolute) difference between the SPT-LS-makespan and the optimum makespan yields

$$\mathbb{E}[C_{max}^{SPT-LS} - C_{max}^*] = \mathbb{E}[C_{max}^{SPT-LS}] - \mathbb{E}[C_{max}^*] \leq$$

$$\leq \frac{n+r}{2(n+1)} - \frac{n+r^2}{2m(n+1)} \xrightarrow{n \to \infty} \frac{1}{2} - \frac{1}{2m}.$$

This result is not that satisfying since our simple calculation yields an upper bound on the expected difference that does not converge to 0 as n tends to infinity. However, we do not intend to provide a more sophisticated analysis of the expected difference. We rather take a look at the ratio of the expected SPT-LS-makespan and the expected optimum makespan instead. For it, we can show

$$\frac{\mathbb{E}[C_{max}^{SPT-LS}]}{\mathbb{E}[C_{max}^*]} \leq \frac{(n+r)(n+m-r)}{2m(n+1)} \frac{2m}{n} = \frac{n^2 - r^2 + m(n+r)}{n(n+1)} =$$

$$= 1 + \frac{m-1}{n+1} + \frac{rm-r^2}{n(n+1)} = 1 + \mathcal{O}(\frac{1}{n}).$$

Thus, for any fixed number m of machines, $\mathbb{E}[C_{max}^{SPT-LS}]$ approaches the (expected) optimum not more slowly than $1 + \mathcal{O}(\frac{1}{n})$ approaches 1 or $\mathcal{O}(\frac{1}{n})$ approaches 0 as n increases. So, the order of convergence is 1.

To complete this section, we compare the expected makespans $\mathbb{E}[C_{max}^{SPT-LS}]$ and $\mathbb{E}[C_{max}^{LPT}]$ in case of two machines. Considering the SPT-LS-rule we know

$$\mathbb{E}[C_{max}^{SPT-LS}] = \begin{cases} \frac{n}{4} + \frac{1}{4} & \text{if } 2 \nmid n, \\ \frac{n}{4} + \frac{n}{4(n+1)} & \text{if } 2 \mid n. \end{cases}$$

Regarding the same stochastic model, Coffman, Frederickson, and Lueker [CFL84] proved that the expected makespan of the LPT-schedule is bounded by

$$\frac{n}{4} + \frac{1}{4(n+1)} \leq \mathbb{E}[C_{max}^{LPT}] \leq \frac{n}{4} + \frac{e}{2(n+1)}.$$

Thus, as n increases, $\mathbb{E}[C_{max}^{LPT}]$ approaches the optimum makespan not more slowly than $1 + \mathcal{O}(\frac{1}{n^2})$ approaches 1. So, the order of convergence of the

LPT-rule is 2 in case $m = 2$. This is twice the order of convergence of the SPT-LS-rule.

Regarding the same stochastic model, we can easily improve the upper bound on the expected makespan of the LPT-schedule in case of two machines. The improvement is due to two facts:

- LPT dominates RLPT in terms of makespan-minimization [CS76] and

- the expected makespan of the RLPT-schedule is known [CFL84] to be exactly

$$\mathbb{E}[C_{max}^{RLPT}] = \frac{n}{4} + \frac{1}{2(n + 1)}.$$

Hence, the expected makespan of the LPT-schedule can be bounded from above by

$$\mathbb{E}[C_{max}^{LPT}] \leq \frac{n}{4} + \frac{1}{2(n + 1)}.$$

As presented before, we are able to compute the expected makespan of the SPT-LS-schedule precisely even in case $m \geq 3$. In contrast to this, we do only have bounds on the expected makespan of the LPT-schedule. Therefore, we made some experimental studies to contribute to this issue in case of two machines. In this case, C_{max} and C_Δ are interconnected in the following way

$$C_{max} = \frac{1}{2} \sum_{j=1}^{n} t_j + \frac{1}{2} C_\Delta.$$

In our experiments, we determined the average C_Δ^{LPT}-value of 10^a independent $(2, n)$-instances. Here, the exponent a takes reasonable integral values. Note that the expected C_Δ-value of the SPT-LS-rule is

$$\mathbb{E}[C_\Delta^{SPT-LS}] = \begin{cases} \frac{1}{2} & \text{if } 2 \nmid n, \\ \frac{n}{2(n+1)} & \text{if } 2 \mid n, \end{cases}$$

and the expected C_Δ-value of the LPT-rule is

$$\frac{0.5}{n + 1} \leq \mathbb{E}[C_\Delta^{LPT}] \leq \frac{1}{n + 1}.$$

Using the $\frac{c}{n+1}$-notation ($c \in \mathbb{R}$) of $\mathbb{E}[C_\Delta^{LPT}]$, Table 2.1 contains the corresponding results rounded to three decimal places due to the number of simulations.

a	n	avg. $C_\Delta^{LPT} \times (n+1)$	a	n	avg. $C_\Delta^{LPT} \times (n+1)$
6	3	1.000	6	4	0.833
6	5	0.890	6	6	0.792
6	9	0.875	6	10	0.791
6	19	0.878	6	20	0.792
6	49	0.877	6	50	0.789
6	99	0.879	6	100	0.791
6	499	0.876	6	500	0.790
6	999	0.876	6	1000	0.790
5	9999	0.873	5	10000	0.789
5	99999	0.870	5	100000	0.790

Table 2.1: Average value of $(n+1)C_\Delta^{LPT}$ for odd and even n

Considering the entries in Table 2.1, it is obvious that the expectation of C_Δ^{LPT} depends on whether the number of jobs is even or odd. Furthermore, n-independent constants $c_{even}, c_{odd} \in \mathbb{R}_{>0}$ so that

$$\mathbb{E}[C_\Delta^{LPT}] = \begin{cases} \frac{c_{odd}}{n+1} & \text{if } 2 \nmid n, \\ \frac{c_{even}}{n+1} & \text{if } 2 \mid n \end{cases}$$

is true for $n \geq 3$ do not seem to exist. However, some more simulations (the results are omitted here) with even larger n (but less number of instances) induced us to the very vague conjecture:

Conjecture 2.5.20
When n tends to infinity, the expectation of C_Δ^{LPT} converges to

$$\lim_{n \to \infty} \mathbb{E}[C_\Delta^{LPT}] = \begin{cases} \frac{\sin \frac{\pi}{3}}{n+1} & \text{if } 2 \nmid n, \\ \frac{\frac{\pi}{4}}{n+1} & \text{if } 2 \mid n. \end{cases}$$

Concerning the difference between the expected makespan of the SPT-LS-schedule and the expected makespan of the LPT-schedule in case of two machines, we finally get the theoretical result (using bounds from [CFL84] and the expected makespan of the SPT-LS-schedule given on page 48):

$$\frac{1}{4} - \frac{3}{n+1} \leq \mathbb{E}[C_{max}^{SPT-LS}] - \mathbb{E}[C_{max}^{LPT}] \leq \frac{1}{4} - \frac{0.25}{n+1}.$$

Thus, as n tends to infinity, the absolute difference between the SPT-LS-makespan and the LPT-makespan is approximately $\frac{1}{4}$ in expectation. This corresponds well with the result on the expected absolute error (C_{max}^{SPT-LS} − C_{max}^*) (cf. page 48), as LPT converges to absolute optimality in expectation [FR86], [FR87].

2.6 LPT-Based Heuristics

This section is concerned with the investigation of LPT-based heuristics such as the RLPT-heuristic which has already been mentioned in Section 2.2 on page 20. We begin with a short introduction to RLPT in case of at least two machines. Then, we do a detailed study on the performance of RLPT for the three specific Identical Machine Scheduling Problems (IMSP) that are relevant to this thesis and compare it with SPT-LS and particularly with LPT. Afterwards, a generalization of LPT and RLPT is suggested and discussed briefly.

2.6.1 The RLPT-Heuristic

Unlike the LPT-heuristic, the RLPT-heuristic (**restricted longest processing time**) generates cardinality-balanced schedules and works as follows: Assign the ranks in increasing order, jobs within a rank in increasing order of the job-indices onto distinct machines as the machines become available after executing all previous ranks. This means that job $km + j$ ($k = 0, \ldots, \frac{n}{m} - 1; j = 1, \ldots, m$) is assigned to the machine with j-th shortest completion time after the assignment of km jobs. Tie situations are broken arbitrarily.

Thus, with the RLPT-heuristic the assignment of the jobs of a certain rank is related to the current machine completion times after executing all previous ranks. This is the main difference compared to the LPT-rule where each job is assigned to the machine with minimum completion time so far, i. e., after the assignment of all previous jobs. So, even within a rank, jobs do not have to be assigned to distinct machines in the LPT-schedule.

Example 2.6.1
Consider the $(3, 6)$-instance with processing times $t_1 = t_2 = 4, t_3 = \ldots t_6 = 1$. Then, RLPT generates the schedule $S^{RLPT} = (1, 2, 3, 1, 2)$ (by application of the tie-break rule of page 15), and LPT generates the schedule $S^{LPT} = (1, 2, 3, 3, 3, 3)$.

We want to mention that an RLPT-schedule can easily be transformed into an SPT-schedule (cf. page 24 for the broad class of SPT-schedules). The transformation is quite intuitive and consists in reversing the execution-order of the jobs on each machine in the RLPT-schedule. We call the result of the transformation *reversed* RLPT-schedule. Note that the transformation does not influence the objective function values of C_{max}, C_{min} and C_Δ. Moreover, the reversed RLPT-schedule also minimizes mean flow time.

The interest in cardinality-balanced schedules arises from practical scheduling problems. Tsai [Tsa92] mentioned two of them, namely the allocation of component types to machines that manufacture VLSI (very large-scale integrated) chips and the assignment of tools to machines in flexible manufacturing systems.

2.6.2 RLPT Compared to SPT-LS

Since RLPT as well as SPT-LS generate cardinality-balanced schedules it is especially interesting if one of the two heuristics is dominated by the other one. The next theorem and the subsequent example deal with this issue.

Theorem 2.6.2
Let I be an arbitrary (m, n)-instance of the IMSP. Then, the following three statements hold:

(i) *The SPT-LS-schedule is at least as long as the RLPT-schedule, i. e.,* $C_{max}^{SPT-LS} \geq C_{max}^{RLPT}$.

(ii) *The C_{min}-value of the SPT-LS-schedule is at most as large as the C_{min}-value of the RLPT-schedule, i. e.,* $C_{min}^{SPT-LS} \leq C_{min}^{RLPT}$.

(iii) *The C_Δ-value of the SPT-LS-schedule is at least as large as the C_Δ-value of the RLPT-schedule, i. e.,* $C_\Delta^{SPT-LS} \geq C_\Delta^{RLPT}$.

Proof
The proof of the three statements is quite simple. As jobs of the same rank have to be assigned to distinct machines in the RLPT-schedule we can conclude that

$$C_{max}^{RLPT} \leq t_1 + t_{m+1} + \ldots + t_{n-m+1} = C_{max}^{SPT-LS}$$

and

$$C_{min}^{RLPT} \geq t_m + t_{2m} + \ldots + t_n = C_{min}^{SPT-LS}$$

which yields

$$C_\Delta^{RLPT} \leq C_\Delta^{SPT-LS}.$$

∎

To verify that SPT-LS is even dominated by RLPT for every integer $m \geq 2$, consider the $(m, 2m)$-instance with processing time $t_j = 2m - j + 1$ for job j ($j = 1, \ldots, 2m$). Then, we can conclude:

- In the RLPT-schedule, each machine has completion time $2m + 1$. Hence, $C_{max}^{RLPT} = C_{min}^{RLPT} = 2m + 1$ which leads to $C_\Delta^{RLPT} = 0$. So, the RLPT-schedule is optimal for all three objectives.

- In the SPT-LS-schedule, the i-th longest running machine has completion time $3m + 2 - 2i$. Hence, $C_{max}^{SPT-LS} = 3m, C_{min}^{SPT-LS} = m + 2$, and $C_\Delta^{SPT-LS} = 2m - 2$.

2.6.3 Main Results on RLPT Compared to LPT

In order to pursue the investigation of LPT- and RLPT-schedules it is useful to take a look at the **profile** of schedules or the set of (current) machine completion times. The ordered m-tuple $a(g) = (a_1(g), \ldots, a_m(g))$ so that $a_i(g) \leq a_{i+1}(g)$ for all i is called *profile after m-group g of schedule S^a*, i. e., after the assignment of the first gm jobs according to algorithm a, if the multiset $\{a_1(g), \ldots, a_m(g)\}$ gives the machine completion times after m-group g. Note that the multisets $\{a_1(g), \ldots, a_m(g)\}$ and $\{C_1^a(gm), \ldots, C_m^a(gm)\}$ are identical. Furthermore, it is important to note that $a_i(g)$ and $a_i(g + 1)$ may correspond to different machines. We also want to mention that $a_1(\frac{n}{m}) = C_{min}^a(n)$, $a_m(\frac{n}{m}) = C_{max}^a(n)$ and $a_m(\frac{n}{m}) - a_1(\frac{n}{m}) = C_\Delta^a(n)$.

Note that the terms *rank* and *m-group* are equivalent for the LPT- and the RLPT-heuristic. We let $l(g)$ denote the LPT-profile and $r(g)$ denote the RLPT-profile after m-group g.

2.6.3.1 The Dominance of LPT over RLPT Concerning C_{min}

In [CS76], Coffman and Sethi proved an interesting result on the makespan of LPT-scheduling compared to RLPT-scheduling.

Theorem 2.6.3 (Coffman and Sethi, 1976)

For each (m, n)-instance the makespan of the RLPT-schedule is not shorter than the makespan of the LPT-schedule, i. e., $C_{max}^{RLPT} \geq C_{max}^{LPT}$.

In particular, this result is interesting as the reversed RLPT-schedule (cf. page 52) is a special SPT-schedule and the shortest SPT-schedule may be up to 20% shorter than the LPT-schedule for a job-system (cf. page 28).

We are especially interested in the question whether LPT is also dominating RLPT concerning the C_{min}-maximization problem or the C_Δ-minimization problem in case $m \geq 3$. Therefore, consider the following theorem.

Theorem 2.6.4

For each (m, n)-instance the minimum completion time of the RLPT-schedule is at most as long as the minimum completion time of the LPT-schedule, i. e., $C_{min}^{RLPT} \leq C_{min}^{LPT}$.

Proof

Remember that $l(g)$ denotes the profile of the LPT-schedule after the assignment of g ranks, and $r(g)$ denotes the profile of the RLPT-schedule after rank g.

In order to prove the theorem we show that for all machines i such that $l_i(g) < r_1(g) + t_{gm}$ it must be true that $l_i(g) \geq r_i(g)$.

As we are particularly interested in the comparison of $l_1(g)$ and $r_1(g)$, with the previous statement we get in case $l_1(g) < r_1(g) + t_{gm}$ that $l_1(g) \geq r_1(g)$ must be true, and in the other case $l_1(g) \geq r_1(g) + t_{gm}$ we can directly conclude that $l_1(g) \geq r_1(g)$ as all processing times are non-negative.

The proof works by induction in the number of ranks.

Base of Induction: $g = 1$

As LPT assigns each job of rank 1 to a different machine, the LPT-schedule and the RLPT-schedule are identical after the assignment of rank 1.

Step of Induction: $g \to g + 1$

Suppose that after group g for all machines i such that $l_i(g) < r_1(g) + t_{gm}$ it is true that $l_i(g) \geq r_i(g)$ and group $(g + 1)$ is being assigned next.

Let $0 \leq k \leq m$ jobs of m-group $(g + 1)$, i. e., the jobs $gm + 1, \ldots, gm + k$ if $k \geq 1$, begin before $r_1(g) + t_{gm}$ in the LPT-schedule. As tasks even within a group do not have to be assigned to distinct machines in the LPT-schedule, these k jobs will be assigned to the first $h \leq k$ elements of the $l(g)$-profile. The remaining $0 \leq m - k \leq m$ jobs of group $(g + 1)$ begin at or after

$r_1(g) + t_{gm}$ in the LPT-schedule. Since $r_1(g + 1) \leq r_1(g) + t_{gm}$, none of the machines that process at least one of the $(m-k)$ shortest jobs of group $(g+1)$ can finish earlier than $r_1(g+1) + t_{(g+1)m}$ in the $l(g+1)$-profile.

In other words, we assume that the following inequality-chain holds in the $l(g)$-profile:

$$r_1(g) + t_{gm} \leq l_1(g) \leq l_2(g) \leq \ldots \leq l_m(g) \text{ (case } k = 0)$$

and

$$l_1(g) \leq l_2(g) \leq \ldots \leq l_h(g) < r_1(g)+t_{gm} \leq l_{h+1}(g) \leq \ldots \leq l_m(g) \text{ (case } k \geq 1).$$

The inductive hypothesis ensures $l_i(g) \geq r_i(g)$ for all $i = 1, \ldots, h$.

Case 1: There exists a machine in the LPT-schedule that processes at least two of the $(m - k)$ shortest jobs of m-group $(g + 1)$.

This directly yields $r_1(g + 1) + t_{(g+1)m} \leq l_1(g + 1)$, i. e., none of the m machines finishes before $r_1(g + 1) + t_{gm+m}$ in the $l(g + 1)$-profile.

Case 2: None of the machines in the LPT-schedule processes more than one of the $(m - k)$ shortest jobs of m-group $(g + 1)$.

Here, we distinguish the following two main subcases $k = h$ and $k > h (\geq 1)$. Each of these two subcases will be subdivided further.

Subcase 1: $k = h$.

In case $h = 0$, each machine processes exactly one job of the current group, i. e., element i of the $l(g)$-profile processes job $(gm + i)$ for $i = 1, \ldots, m$. As none of the jobs of the current group starts before $r_1(g) + t_{gm}$ in the LPT-schedule, we can conclude further that no machine finishes earlier than $r_1(g + 1) + t_{gm+m}$ in the $l(g + 1)$-profile.

In case $h > 0$, element i of the $l(g)$-profile processes job $(gm + i)$ for $i = 1, \ldots, h$. From the inductive hypothesis we know that $l_i(g) + t_{gm+i} \geq r_i(g) + t_{gm+i}$ for $i = 1, \ldots, h$.

As mentioned earlier, the $(m - k)$ machines that process exactly one of the $(m - k)$ shortest jobs of the current group cannot finish earlier than $r_1(g + 1) + t_{gm+m}$.

Provided that the last $(m-k)$ elements of the $l(g)$-profile each process exactly one of the $(m - k)$ shortest jobs of the current group, only the corresponding machines to the first h elements in the $l(g)$-profile can finish earlier than $r_1(g + 1) + t_{gm+m}$ in the $l(g + 1)$-profile. Due to the inductive hypothesis we can find for each of these h machines a distinct machine in the $r(g+1)$-profile which finishes not later. Thus, inequality $l_i(g+1) \geq r_i(g+1)$ holds for all i such that $l_i(g+1) < r_1(g+1) + t_{(g+1)m}$.

In the other case, at least one of the first h elements in the $l(g)$-profile processes one of the $(m-k)$ shortest jobs. So, assume that $\bar{h} \leq \min\{h, m-h\}$ of the first h elements in the $l(g)$-profile process exactly one of the shortest $(m-k)$ jobs. This means that the last $\bar{h}$ elements in the $l(g)$-profile do not process any job of the current group. Further, assume that $l_{h_1}(g) + t_{gm+h_1}$ ($h_1 \in \{1, \dots, h\}$) is the longest current completion time after $(gm+h)$ jobs assigned of all $\bar{h}$ machines of the LPT-schedule that process exactly one of the longest k and one of the shortest $(m-k)$ jobs of the current group. Then, we can conclude: $l_{h_1}(g) + t_{gm+h_1} \leq l_{m-\bar{h}+1}(g)$.

We also know that only the $(h-\bar{h})$ elements out of the first h elements in the $l(g)$-profile which process exactly one job of the current group and the last $\bar{h}$ elements in the $l(g)$-profile which do not process any job of the current group can finish earlier than $r_1(g+1) + t_{gm+m}$ in the $l(g+1)$-profile. So, we can find again for each of these at most h machines a distinct machine in the $r(g+1)$-profile which finishes not later. Thus, inequality $l_i(g+1) \geq r_i(g+1)$ holds for all i such that $l_i(g+1) < r_1(g+1) + t_{(g+1)m}$.

Subcase 2: $k > h$.

In this case, at least one of the first $h \geq 1$ elements in the $l(g)$-profile processes more than one of the k longest jobs of the current group. This means that at least the last $(k-h)$ elements of the $l(g)$-profile do not process any job of the current m-group.

Assume that $(gm+\bar{k})$ $(2 \leq \bar{k} \leq h+1 \leq k)$ is the first job in the current group that is not assigned to element $l_{\bar{k}}(g)$. So, after the assignment of $(gm + \bar{k} - 1)$ jobs we have the current machine completion times

$$l_i(g) + t_{gm+i} \ (i = 1, \dots, \bar{k}-1) \text{ and } l_i(g) \ (i = \bar{k}, \dots, m)$$

in the LPT-schedule and

$$r_i(g) + t_{gm+i} \ (i = 1, \dots, \bar{k}-1) \text{ and } r_i(g) \ (i = \bar{k}, \dots, m)$$

in the RLPT-schedule. As the first h elements in the $l(g)$-profile fulfill the condition $l_i(g) < r_1(g) + t_{gm}$ $(i = 1, \dots, h)$, the inductive hypothesis ensures

$$l_i(g) + t_{gm+i} \geq r_i(g) + t_{gm+i} \text{ for } i = 1, \dots, \bar{k}-1.$$

Job $(gm + \bar{k})$ is assigned to one of the first $(\bar{k} - 1)$ elements in the $l(g)$-profile, i. e.,

$$\min_{1 \leq i \leq \bar{k}-1} \{l_i(g) + t_{gm+i}\} < l_{\bar{k}}(g).$$

In particular, we know

$$r_1(g + 1) \leq \min_{1 \leq i \leq \bar{k}-1} \{r_i(g) + t_{gm+i}\} \leq \min_{1 \leq i \leq \bar{k}-1} \{l_i(g) + t_{gm+i}\}.$$

By this, we can directly conclude that none of the machines that process at least one of the jobs $gm + \bar{k}, \ldots, gm + m$ can finish earlier than $r_1(g + 1) + t_{gm+m}$ in the $l(g + 1)$-profile.

Thus, if a machine processes at least two of the jobs $gm + \bar{k}, \ldots, gm + m$, then none of the m machines finishes before $r_1(g+1) + t_{gm+m}$ in the $l(g + 1)$-profile.[6]

In the other case, i. e., the jobs $gm + \bar{k}, \ldots, gm + m$ are assigned to distinct machines[7], at most the elements out of the first $(\bar{k} - 1)$ elements in the $l(g)$-profile that process exactly one job of the current group and the last $(k - h)$ elements in the $l(g)$-profile which do not process any job of the current group can finish earlier than $r_1(g + 1) + t_{gm+m}$ in the $l(g+1)$-profile. These are at most $(\bar{k}-1-(k-h)) + (k - h) = \bar{k} - 1 \leq h$ machines, and for each of them in the $l(g + 1)$-profile we can find a distinct machine in the $r(g + 1)$-profile which finishes not later. This is correct since $l_i(g) + t_{gm+i} \geq r_i(g) + t_{gm+i}$ for all $i \in \{1, \ldots, \bar{k} - 1\}$ and $l_{m-k+h+1}(g) \geq l_{\bar{k}}(g) > \min_{i=1,\ldots,\bar{k}-1}\{l_i(g) + t_{gm+i}\}$. Thus, we get $l_i(g + 1) \geq r_i(g + 1)$ for all i such that $l_i(g + 1) < r_1(g + 1) + t_{(g+1)m}$.

This completes the proof of Theorem 2.6.4[8]. ∎

The next corollary follows directly from the two results $C_{max}^{LPT} \leq C_{max}^{RLPT}$ and $C_{min}^{LPT} \geq C_{min}^{RLPT}$.

Corollary 2.6.5

For each (m,n)-instance the C_Δ-value of the RLPT-schedule is at least as large as the C_Δ-value of the LPT-schedule, i. e., $C_\Delta^{RLPT} \geq C_\Delta^{LPT}$.

---

[6]Note that this is only possible for the first h elements in the $l(g)$-profile in this subcase.

[7]Note that this is only possible if inequality $k \leq 2(\bar{k} - 1) \leq 2h$ holds in this subcase.

[8]By the proof of Theorem 2.6.4, we have additionally shown that the number of machines i that fulfill $l_i(g) < r_1(g) + t_{gm}$ is monotonically decreasing in the number of groups assigned.

Before moving to a worst-case analysis of the RLPT-heuristic, we take a brief look at the following issue. Assume that $F \in \{C_{max}, C_{min}, C_\Delta\}$ is the objective function under consideration and let $F^A \prec F^B$ denote that algorithm A yields a strictly better (current) result for F than algorithm B. Then, we are interested in the question whether job-systems exist where RLPT, after $F^{LPT}(j) \prec F^{RLPT}(j)$ for some intermediate $j \in \{m+2, \ldots, n-1\}$[9], can return to $F^{LPT}(k) = F^{RLPT}(k)$ for some $k \in \{j+1, \ldots, n\}$. Especially, we are interested in the smallest possible (or minimal) value for k.

- $F = C_{max}$.
  While at least $(m+2)$ jobs have to be assigned so that currently $C_{max}^{LPT} < C_{max}^{RLPT}$ can hold for the first time, at least $(m+3)$ jobs have to be assigned so that RLPT can return to $C_{max}^{LPT} = C_{max}^{RLPT}$. Hence, the following $(m, m+3)$-instance is a minimal example for any fixed $m \geq 2$. This means that $k = m+3$ is minimal concerning C_{max}. The job-system is:

$$t_1 = \ldots = t_{m-1} = 4,$$
$$t_m = 2,$$
$$t_{m+1} = t_{m+2} = t_{m+3} = 1.$$

The relevant objective function values contained in the subsequent table are readily verified.

j	$m+1$	$m+2$	$m+3$
$C_{max}^{LPT}(j)$	4	4	5
$C_{max}^{RLPT}(j)$	4	5	5

- $F = C_{min}$.
  Due to the facts that $C_{min}^{LPT}(m+1) = C_{min}^{RLPT}(m+1)$ and LPT - in contrast to RLPT - always chooses a machine with current minimum completion time, at least $(2m+1)$ jobs have to be assigned so that RLPT, after $C_{min}^{LPT}(j) > C_{min}^{RLPT}(j)$ for some intermediate $j \in \{m+2, \ldots, 2m\}$, can return to $C_{min}^{LPT} = C_{min}^{RLPT}$.
  By adding $(m-2)$ length 1 jobs to the previous $(m, m+3)$-instance we get an $(m, 2m+1)$-instance which is minimal for $F = C_{min}$ and any fixed $m \geq 2$. So, $k = 2m+1$ is minimal concerning $F = C_{min}$. The

---

[9]Note that $F^{LPT}(j) = F^{RLPT}(j)$ is valid for $j = 1, \ldots, m+1$ and all three F under consideration.

job-system is:

$$t_1 = \ldots = t_{m-1} = 4,$$
$$t_m = 2,$$
$$t_{m+1} = \ldots = t_{2m+1} = 1.$$

Again, the relevant objective function values are contained in the subsequent table.

j	$m+1$	$m+2$	$\ldots$	$2m$	$2m+1$
$C_{min}^{LPT}(j)$	3	4	$\ldots$	4	4
$C_{min}^{RLPT}(j)$	3	3	$\ldots$	3	4

- $F = C_\Delta$.

  For any (m,n)-instance we know that equality $C_\Delta^{LPT} = C_\Delta^{RLPT}$ can only be achieved if $C_{max}^{LPT} = C_{max}^{RLPT}$ and $C_{min}^{LPT} = C_{min}^{RLPT}$ hold. Thus, the previous $(m, 2m+1)$-instance is also minimal for $F = C_\Delta$, i. e., $k = 2m + 1$ is minimal concerning $F = C_\Delta$ and any fixed $m \geq 2$.

2.6.3.2 Worst-Case Analysis of RLPT

This part deals with a worst-case analysis of the RLPT-heuristic. At first, the C_{max}-minimization problem is considered and afterwards the C_{min}-maximization problem.

So, let us start with the objective function C_{max}. We will precisely analyze the worst-case performance of the RLPT-heuristic for it. By Theorem 2.6.2 of page 52 we know that inequality $C_{max}^{RLPT} \leq C_{max}^{SPT-LS}$ holds for each (m,n)-instance. This yields

$$\frac{C_{max}^{RLPT}}{C_{max}^*} \leq \frac{C_{max}^{SPT-LS}}{C_{max}^*} \leq 2 - \frac{1}{m}.$$

Furthermore, we know that the reversed RLPT-schedule belongs to the broad class of SPT-schedules. So, for each (m,n)-instance the inequality $C_{max}^{RLPT} \geq C_{max}^{SPT^*}$ holds. Remember that SPT* refers to a schedule with shortest maximum completion time among all SPT-schedules. In Subsection 2.4.2 we have already mentioned the bounds

$$\frac{C_{max}^{SPT^*}}{C_{max}^*} \leq 2 - \frac{1}{m} \quad \text{and} \quad \frac{C_{max}^{SPT^*}}{C_{max}^{LPT}} \leq 2 - \frac{1}{m}$$

which were presented by Bruno, Coffman, and Sethi in [BCS74]. The authors also give a family of instances so that the ratio $\frac{C_{max}^{SPT^*}}{C_{max}^*}$ is asymptotically $(2 - \frac{1}{m})$.

Due to the inequalities $C_{max}^{SPT^*} \leq C_{max}^{RLPT} \leq C_{max}^{SPT-LS}$ we know that the performance bound

$$\frac{C_{max}^{RLPT}}{C_{max}^*} \leq 2 - \frac{1}{m}$$

is asymptotically tight.

With the next theorem we show that no instance exists so that the makespan of the RLPT-schedule is exactly $(2 - \frac{1}{m})$ times the optimum makespan or the makespan of the LPT-schedule.

Theorem 2.6.6

The performance bounds

$$\frac{C_{max}^{RLPT}}{C_{max}^*} \leq 2 - \frac{1}{m} \quad and \quad \frac{C_{max}^{RLPT}}{C_{max}^{LPT}} \leq 2 - \frac{1}{m}$$

are asymptotically tight for any fixed number m of machines but cannot be reached exactly.

Note that the same performance bound $(2 - \frac{1}{m})$ applies when LPT-scheduling is considered instead of optimum scheduling, and that the bound is still best possible.

Proof

We prove the inequality

$$\frac{C_{max}^{RLPT}}{C_{max}^*} < 2 - \frac{1}{m}$$

which directly leads to

$$\frac{C_{max}^{RLPT}}{C_{max}^{LPT}} < 2 - \frac{1}{m}$$

as $C_{max}^{LPT} \geq C_{max}^*$. To verify that both bounds are asymptotically tight for any fixed number m, we give a family of job-systems at the end of this proof. To prove the first inequality, we use the Theorems 2.1.2 (cf. page 15) and 2.6.2 (cf. page 52). As the SPT-LS-schedule is at least as long as the RLPT-schedule we take a look at such job-systems for which the makespan of the SPT-LS-schedule is exactly $(2 - \frac{1}{m})$ times the optimal makespan. We will see that RLPT generates shorter schedules than SPT-LS for those job-systems. Therefore, we apply Theorem 2.1.2 to SPT-LS. In order to generate an LS-schedule whose makespan is $(2 - \frac{1}{m})$ times the optimal makespan, each inequality in the proof of Theorem 2.1.2 has to be a strict equality. Hence, the following three properties have to be fulfilled:

(i) Before assigning the last job j on the makespan-machine, each of the m machines must have the same completion time so far.

(ii) The length of job j equals the optimal makespan.

(iii) The weighted sum of all processing times equals the optimal makespan, i. e., $\frac{1}{m} \sum_{k=1}^{n} t_k = C_{max}^*$.

According to the SPT-LS-rule we know by Theorem 2.4.1 (cf. page 25) that the machine which processes the job J_1 is a makespan-machine. As this job is the last one in the list we can conclude that the job-system must contain $(m-1)$ dummy-jobs because otherwise not all m machines can have the same current completion time before the last job is assigned by SPT-LS. Moreover, the processing times must fulfill the following relation:

$$C_{max}^* = t_1 \geq t_2 = \ldots = t_{m+1} \geq$$
$$\geq t_{m+2} = \ldots = t_{2m+1} \geq$$
$$\vdots$$
$$\geq t_{n-2m+2} = \ldots = t_{n-m+1} >$$
$$> t_{n-m+2} = \ldots = t_n = 0.$$

By the properties (ii) and (iii) from above we can conclude further that

$$t_1 > t_2 = \ldots = t_{m+1} > 0.$$

Regarding job-systems in which the processing times fulfill the previous relations, we distinguish two cases in order to compare the makespan produced by RLPT-scheduling with the makespan produced by SPT-LS-scheduling.

Case 1: $t_2 = \ldots = t_{n-m+1} > 0$.

Then, we can conclude:

$$C_{max}^{RLPT} = \max \left\{ t_1 + \left(\frac{n}{m} - 2 \right) t_2, \frac{n}{m} t_2 \right\}.$$

As $t_1 + (\frac{n}{m} - 2)t_2 < t_1 + (\frac{n}{m} - 1)t_2$ and $\frac{n}{m}t_2 < t_1 + (\frac{n}{m} - 1)t_2$, we get

$$C_{max}^{RLPT} < t_1 + (\frac{n}{m} - 1)t_2 = C_{max}^{SPT-LS}.$$

Case 2: $t_2 = \ldots = t_{km+1} > t_{km+2} > 0$ for some $k \in \{1, \ldots, \frac{n}{m} - 2\}$.

Then, we can conclude that

$$C_{max}^{RLPT}(km) = t_1 + t_{m+1} + \ldots + t_{(k-1)m+1},$$

whereas all other machines have current completion time kt_2 after the assignment of km jobs. Hence,

$$C_{max}^{RLPT}(km + m) < t_1 + t_{m+1} + \ldots + t_{(k-1)m+1} + t_{km+1}$$

which leads to

$$C_{max}^{RLPT} < t_1 + t_{m+1} + \ldots + t_{n-m+1} = C_{max}^{SPT-LS}.$$

To sum up, RLPT generates shorter schedules than SPT-LS in both cases. This yields

$$\frac{C_{max}^{RLPT}}{C_{max}^{*}} < 2 - \frac{1}{m} \quad \text{and} \quad \frac{C_{max}^{RLPT}}{C_{max}^{LPT}} < 2 - \frac{1}{m}$$

since $C_{max}^{LPT} \geq C_{max}^{*}$.

To verify that these bounds are asymptotically tight, consider the following job-system and assume $d \in \mathbb{N}$:

$$t_1 = dm, t_2 = \ldots = t_{dm(m-1)+1} = 1.$$

Then, we get $C_{max}^{*} = dm = C_{max}^{LPT}$, whereas the makespan produced by RLPT is

$$C_{max}^{RLPT} = dm + (dm - d - 1) = 2dm - d - 1.$$

Thus, the ratio is

$$\frac{2dm - d - 1}{dm} = 2 - \frac{1}{m} - \frac{1}{dm} \xrightarrow[d \to \infty]{} 2 - \frac{1}{m}.$$

∎

To complete the analysis concerning the objective function C_{max} we state the obvious lower bound on $\frac{C_{max}^{RLPT}}{C_{max}^{*}}$ and the known lower bound on $\frac{C_{max}^{RLPT}}{C_{max}^{LPT}}$ [CS76]:

$$1 \leq \frac{C_{max}^{RLPT}}{C_{max}^{*}} \quad \text{and} \quad 1 \leq \frac{C_{max}^{RLPT}}{C_{max}^{LPT}}.$$

Again, it is interesting to note that the same lower bound applies when LPT-scheduling is considered instead of optimal scheduling.

Next, we briefly deal with the worst-case performance of RLPT-scheduling compared to LPT-scheduling as well as optimal scheduling concerning the problem of maximizing C_{min}. By Woeginger's result [Woe97] (cf. page 17) and statement (ii) of Theorem 2.6.2 we can conclude:

$$\frac{1}{m} \leq \frac{C_{min}^{RLPT}}{C_{min}^{*}}.$$

Furthermore, Theorem 2.6.4 assures $C_{min}^{RLPT} \leq C_{min}^{LPT}$. The following theorem can be proved analogously to the proof of Theorem 2.6.6.

Theorem 2.6.7
The performance bounds

$$\frac{C_{min}^{RLPT}}{C_{min}^{*}} \geq \frac{1}{m} \quad and \quad \frac{C_{min}^{RLPT}}{C_{min}^{LPT}} \geq \frac{1}{m}$$

are asymptotically tight for any fixed number m of machines but cannot be reached exactly.

Proof
In an analogous manner to the proof of Theorem 2.6.6, we can prove Theorem 2.6.7. For this reason, the proof is omitted here. ∎

We rather present a family of job-systems for any fixed number $m \geq 2$ that approaches the bound. Therefore, assume $d \in \mathbb{N}$ and consider the following job-system:

$$t_1 = \ldots = t_{m-1} = dm,$$
$$t_m = \ldots t_{dm+m-1} = 1.$$

Then, we get $C_{min}^{*} = dm = C_{min}^{LPT}$, whereas the minimum completion time of the RLPT-schedule is $C_{min}^{RLPT} = d + 1$. Hence,

$$\frac{C_{min}^{RLPT}}{C_{min}^{*}} = \frac{C_{min}^{RLPT}}{C_{min}^{LPT}} = \frac{d+1}{dm} = \frac{1}{m} + \frac{1}{dm} \xrightarrow{d \to \infty} \frac{1}{m}.$$

Again, it is interesting to note that the same bounds apply when RLPT-scheduling is compared to LPT-scheduling instead of optimal scheduling. Moreover, let SPT_{min}^{*} refer to an SPT-schedule with longest minimum completion time among all SPT-schedules. Then, the previous job-system (added by $t_{dm+m} = 0$) reveals that the bounds

$$\frac{C_{min}^{SPT_{min}^{*}}}{C_{min}^{*}} \geq \frac{1}{m} \quad and \quad \frac{C_{min}^{SPT_{min}^{*}}}{C_{min}^{LPT}} \geq \frac{1}{m}$$

are asymptotically tight, too. However, it is not true that $C_{min}^{SPT^*_{min}} \leq C_{min}^{LPT}$ is valid for all instances. This means that instances exist so that $C_{min}^{SPT^*_{min}} > C_{min}^{LPT}$. We show this by the following $(m, 3m)$-instance which contains one dummy-job:

$$
\begin{aligned}
t_j &= 3m - 1 - j, & j &= 1, \ldots, m, \\
t_j &= 3m - j, & j &= m + 1, \ldots, 2m, \\
t_j &= m, & j &= 2m + 1, \ldots, 3m - 1, \\
t_{3m} &= 0.
\end{aligned}
$$

Then, it is readily verified that $C_{min}^{LPT} = 4m - 2$, whereas we get $C_{min}^{SPT^*_{min}} = 5m - 3$. This means that the minimum completion time of a best SPT-schedule concerning the objective function C_{min} can be (at least) up to 25% longer than the minimum completion time of the LPT-schedule.

Coffman's theoretical result [Cof73] on the performance ratio $\frac{C_{max}^{SPT^*}}{C_{max}^{LPT}}$ and the corresponding job-system he presented are quite similar to our result. So, we are led to the following (vague) conjecture.

Conjecture 2.6.8

The performance ratio of $C_{min}^{SPT^*_{min}}$ and C_{min}^{LPT} is

$$
\frac{C_{min}^{SPT^*_{min}}}{C_{min}^{LPT}} \leq \frac{5m - 3}{4m - 2}
$$

for any fixed number m of machines.

Finally, we mention the obvious upper bound on $\frac{C_{min}^{RLPT}}{C_{min}^*}$ and the upper bound on $\frac{C_{min}^{RLPT}}{C_{min}^{LPT}}$ which is given by Theorem 2.6.4:

$$
\frac{C_{min}^{RLPT}}{C_{min}^*} \leq 1 \quad \text{and} \quad \frac{C_{min}^{RLPT}}{C_{min}^{LPT}} \leq 1.
$$

Again, it is interesting to note that the same upper bound applies when LPT-scheduling is considered instead of optimal scheduling.

2.6.3.3   Experimental Results in Case of Two Machines

The theoretical results concerning the comparison of LPT- and RLPT-scheduling have shown that RLPT-scheduling cannot lead to better objective function values than LPT-scheduling concerning any of the three objective functions that are considered in this thesis. In this context, an interesting question is: How often do the LPT- and the RLPT-heuristic lead to schedules with the same objective function value (or even to identical schedules)? In the sequel, we present some results concerning this question in case of two machines. We took the stochastic model of Section 2.5.2 (cf. page 47) and determined experimentally the probability that the LPT- and the RLPT-heuristic generate the same schedule for a random $(2, n)$-instance. Remember that we assume the n processing times to be independent samples, uniformly distributed in the unit interval $[0, 1]$ and that n denotes the number of non-dummy-jobs in our experiments.

Our experimental results (cf. Table 2.2) led us to the following conjecture.

Conjecture 2.6.9

Assume the processing times to be independent samples, uniformly distributed in the unit interval $[0, 1]$ and assume $m = 2$ and $n \geq 4$. Then, the probability $Pr\{S^{LPT}(1:n) = S^{RLPT}(1:n)\}$ that the LPT- and the RLPT-heuristic generate the same schedule is

$$Pr\{S^{LPT}(1:n) = S^{RLPT}(1:n)\} = \begin{cases} \frac{3}{4} & \text{if } 2 \mid n, \\ \frac{7}{8} & \text{if } 2 \nmid n. \end{cases}$$

The next table contains some experimental results that support Conjecture 2.6.9. For each number of jobs we generated 10^7 independent $(2, n)$-instances. The entries are rounded to three decimal places.

n	4	5	6	7	8	9
$S^{LPT} = S^{RLPT}$	74.981	87.501	74.992	87.506	75.021	87.520

n	10	11	12	13	24	25
$S^{LPT} = S^{RLPT}$	75.010	87.488	75.021	87.484	75.000	87.489

n	50	51	52	53	100	101
$S^{LPT} = S^{RLPT}$	74.998	87.503	74.984	87.497	74.994	87.489

n	102	103	250	251	500	501
$S^{LPT} = S^{RLPT}$	75.008	87.491	74.990	87.496	74.998	87.481

Table 2.2: Relative number (in %) of instances so that the LPT- and the RLPT-heuristic generate the same schedule

Before we state two more results, we focus on the question whether the following statement is equivalent to Conjecture 2.6.9 (we assume the same conditions as in Conjecture 2.6.9):

Statement 2.6.10

The probability $Pr\{C_{max}^{LPT}(n) < C_{max}^{RLPT}(n)\}$ *that the LPT-schedule is shorter than the RLPT-schedule is*

$$Pr\{C_{max}^{LPT}(n) < C_{max}^{RLPT}(n)\} = \begin{cases} \frac{1}{4} & \text{if } 2 \mid n, \\ \frac{1}{8} & \text{if } 2 \nmid n. \end{cases}$$

At first, note that even in case of two machines it is generally not correct to conclude from different schedules S^{LPT} and S^{RLPT} to different makespans (cf. item $F = C_{max}$ on page 58). Nevertheless, the question whether Conjecture 2.6.9 and Statement 2.6.10 are equivalent will be answered positively. Remember that we assume the longest job to be assigned to machine 1, and note that the event $C_{max}^A(j) = C_{min}^A(j)$ $(A \in \{LPT, RLPT\})$ has probability zero for all $j = 1, \ldots, n$ since the processing times are drawn from a continuous probability distribution. This is still not enough to conclude from different schedules S^{LPT} and S^{RLPT} to different makespans. So, it is useful to take a look at the (rare) cases where RLPT, after $C_{max}^{LPT}(j) < C_{max}^{RLPT}(j)$ for some intermediate $j < n$, returns to $C_{max}^{LPT}(n) = C_{max}^{RLPT}(n)$ at the end. The following two lemmas are meant to classify such cases. Thereby, j^* denotes the job after whose assignment $C_{max}^{LPT}(j^*) < C_{max}^{RLPT}(j^*)$ holds for the first time (if such a job exists in a given $(2,n)$-instance). It is readily verified that j^* is even and $j^* \geq 4$, if such a job j^* exists.[10] The proofs are to be found in the Appendix A.2.

Lemma 2.6.11

Consider an arbitrary $(2,n)$-instance that contains a job j^* $(j^* < n)$. If $C_{max}^{RLPT}(2j) - C_{min}^{LPT}(2j) > t_{2j}$ for some $j \in \{\frac{j^*}{2}, \ldots, \frac{n}{2} - 1\}$, then

(i) $C_{max}^{LPT}(k) < C_{max}^{RLPT}(k)$ *for all* $k = 2j + 1, \ldots, n$ *and*

(ii) $C_{max}^{RLPT}(2k) - C_{min}^{LPT}(2k) > t_{2k}$ *for all* $k = \frac{j}{2} + 1, \ldots, \frac{n}{2}$.

By Lemma 2.6.11 and the Lemmas A.1.4 and A.1.5 of Appendix A.1 we get the necessary condition $C_{max}^{RLPT}(2j) - C_{min}^{LPT}(2j) = t_{2j}$ for all $j \in \{\frac{j^*}{2}, \ldots, \frac{n}{2} - 1\}$ in order to obtain equally good schedules. The next lemma deals with this condition.

Lemma 2.6.12

Consider an arbitrary $(2,n)$-instance that contains a job j^* $(j^* < n)$. Furthermore, let $C_{max}^{RLPT}(2j) - C_{min}^{LPT}(2j) = t_{2j}$ for some $j \in \{\frac{j^*}{2}, \ldots, \frac{n}{2} - 1\}$.

---

[10]A necessary condition for the existence of j^* is that the given job-system contains at least two different positive processing times.

(i) If $t_{2j} > t_{2j+1}$, then $C_{max}^{LPT}(k) < C_{max}^{RLPT}(k)$ for all $k = 2j + 1, \ldots, n$.

(ii) If $t_{2j} = t_{2j+1}$, then $C_{max}^{LPT}(2j + 1) = C_{max}^{RLPT}(2j + 1)$.

(iii) If $t_{2j} = t_{2j+1} = t_{2j+2}$, then $C_{max}^{LPT}(2j + 2) < C_{max}^{RLPT}(2j + 2)$ and
$C_{max}^{RLPT}(2j + 2) - C_{min}^{LPT}(2j + 2) = t_{2j+2}$.

(iv) If $t_{2j} = t_{2j+1} > t_{2j+2} > 0$, then $C_{max}^{LPT}(2j + 2) < C_{max}^{RLPT}(2j + 2)$ and

$$C_{max}^{RLPT}(2j+2) - C_{min}^{LPT}(2j+2) \begin{cases} = t_{2j+2} & \text{if } C_{min}^{LPT}(2j + 2) = C_{max}^{LPT}(2j + 1), \\ > t_{2j+2} & \text{if } C_{min}^{LPT}(2j + 2) < C_{max}^{LPT}(2j + 1). \end{cases}$$

Statement (i) of Lemma 2.6.12 reveals that the condition $t_{2j} = t_{2j+1}$ for all $j \in \{\frac{j^*}{2}, \ldots, \frac{n}{2} - 1\}$ is necessary, too. The event that two processing times are equal has probability zero in case of a continuous probability distribution. Thus, Statement 2.6.10 is equivalent to Conjecture 2.6.9.

Apart from this, Lemma 2.6.12 reveals the following: If job j^* exists in a given $(2, n)$-instance that contains an even number of jobs with positive lengths, then LPT-scheduling leads to a shorter schedule than RLPT-scheduling.

For the same stochastic model we were also interested in the question: What is the probability that LPT- and RLPT-scheduling generate the same (partial) schedule after the assignment of the longest $4 \le j \le n$ jobs. Here, our experimental results (cf. Appendix A.3) led us to the following conjecture.

Conjecture 2.6.13
Assume the processing times to be independent samples, uniformly distributed in the unit interval $[0, 1]$ and assume $m = 2$ and $n \ge 4$. Then, the probability $Pr\{S^{LPT}(1 : n - k) = S^{RLPT}(1 : n - k)\}$ that the LPT- and the RLPT-heuristic generate the same schedule after the assignment of the $(n-k)$ longest jobs is

$$Pr\{S^{LPT}(1 : n - k) = S^{RLPT}(1 : n - k)\} = 1 - \frac{1}{2^{k+2}}$$

in case n and k are of same parity and $n \ge k + 4$.

Note that the two cases $k = 0$ (and n even) and $k = 1$ (and n odd) are consistent with Conjecture 2.6.9.

Although we do not have a theoretical proof of the previous two Conjectures 2.6.9 and 2.6.13, we are able to prove a statement which is related to Conjecture 2.6.13.

Theorem 2.6.14

Assume the processing times to be independent samples, uniformly distributed in the unit interval $[0, 1]$ and assume $m = 2$ and $n \geq 4$. Then, the probability $Pr\{S^{LPT}(1 : 4) = S^{RLPT}(1 : 4)\}$ that the LPT- and the RLPT-heuristic generate the same schedule after the assignment of the four longest jobs is

$$Pr\{S^{LPT}(1 : 4) = S^{RLPT}(1 : 4)\} = 1 - \frac{1}{2^{n-2}}.$$

So, the statement of Theorem 2.6.14 is a special case of Conjecture 2.6.13 with $k = n - 4$.

Proof

To prove Theorem 2.6.14 we need some information on the joint density function of three order statistics. The event that the LPT- and the RLPT-heuristic generate the same schedule after the assignment of the four longest jobs is equivalent to the event that the sum of the second largest and the third largest order static is greater than the largest order statistic. Hence, we can determine the probability $Pr\{X_{1:n} < X_{2:n} + X_{3:n}\}$ to prove the theorem. Therefore, we need the joint density function $f_{3,2,1:n}$ of the third, the second and the largest order statistic. The joint density function of three order statistics can be found in [ABN92, pages 25-26]. We get

$$f_{3,2,1:n}(x_3, x_2, x_1) = \frac{n!}{(n-3)!} x_3^{n-3}, \quad (0 \leq x_3 < x_2 < x_1 \leq 1),$$

in our stochastic model. Then, the desired probability can be calculated as follows:

$$Pr\{X_{1:n} < X_{2:n} + X_{3:n}\} =$$

$$= \int_{x_3=0}^{1} \int_{x_2=x_3}^{1} \int_{x_1=x_2}^{1} f_{3,2,1:n}(x_3, x_2, x_1) \mathbf{1}_{(x_1 < x_2 + x_3)} dx_1 dx_2 dx_3 =$$

$$= \int_{0}^{1} \int_{x_3}^{1} \int_{x_2}^{\min\{1, x_2+x_3\}} \frac{n!}{(n-3)!} x_3^{n-3} dx_1 dx_2 dx_3 =$$

$$= \frac{n!}{(n-3)!} \left[\int_{0}^{\frac{1}{2}} \int_{x_3}^{1-x_3} \int_{x_2}^{x_2+x_3} x_3^{n-3} dx_1 dx_2 dx_3 + \right.$$

$$\left. + \int_{0}^{\frac{1}{2}} \int_{1-x_3}^{1} \int_{x_2}^{1} x_3^{n-3} dx_1 dx_2 dx_3 + \int_{\frac{1}{2}}^{1} \int_{x_3}^{1} \int_{x_2}^{1} x_3^{n-3} dx_1 dx_2 dx_3 \right] =$$

$$= \frac{n!}{(n-3)!} \left[\frac{1}{n(n-1)} \frac{1}{2^{n-1}} + \frac{1}{n} \frac{1}{2^{n+1}} + \right.$$

$$+ \frac{1}{n(n-1)(n-2)} - \frac{1}{(n-1)(n-2)} \frac{1}{2^{n-1}} - \frac{1}{n} \frac{1}{2^{n+1}} \right] =$$

$$= \frac{n!}{(n-3)!} \left[\frac{1}{n(n-1)(n-2)} - \frac{2}{n(n-1)(n-2)} \frac{1}{2^{n-1}} \right] = 1 - \frac{1}{2^{n-2}}.$$

∎

We also made some experiments for $m \geq 3$ as well as for other distributions of the processing times (e. g. exponential distribution). However, here we did not get similarly nice results.

2.6.4 The k-LPT-Heuristics

As a generalization of the two heuristics LPT and RLPT we propose the k-LPT-heuristics ($k \in \{1, \ldots, m\}$). The main idea is to assign k jobs at a time, each of them onto a distinct machine. More precisely, the k-LPT-heuristic works as follows.

At first, the jobs are sorted in non-increasing order of processing times. Then, the jobs are partitioned into $\lceil \frac{n}{k} \rceil$ k-groups with jobs $J_{gk+1}, \ldots, J_{gk+k}$ in k-group $(g+1)$ for $g = 0, 1, \ldots, \lceil \frac{n}{k} \rceil - 2$ and jobs $J_{\lceil \frac{n}{k} \rceil k - k + 1}, \ldots, J_n$ in the last group which consists of at most k jobs.

Assign the k-groups in increasing order, jobs within a k-group in increasing order of the job-indices onto distinct machines as the machines become available after executing all previous k-groups. This means that job $gk + j$ ($g = 0, \ldots, \lceil \frac{n}{k} \rceil - 1; j = 1, \ldots, k$[11]) is assigned to the machine with j-th shortest completion time after the assignment of gk jobs.

Note that the LPT-heuristic corresponds to $k = 1$ and the RLPT-heuristic corresponds to $k = m$, respectively.

Take care that in contrast to Section 2.5.1 (cf. page 30) the term k-group is related to the processing times of the jobs and not to a given priority list. Especially, in case $k = m$ the terms k-group and rank are equivalent here.

Example 2.6.15
Let a $(4, 10)$-instance be given and let $k = 3$. Then, the 10 jobs are partitioned into three groups of size 3 and one group of size 1. The first k-group contains the jobs J_1, J_2, J_3, the second k-group contains the jobs J_4, J_5, J_6, the third k-group contains the jobs J_7, J_8, J_9, and the last k-group contains job J_{10}.

---

[11]Regarding the last k-group, the variable j takes at most k different values.

In the remainder of this subsection assume $m \geq 3$. The issue that we briefly discuss here arises from the comparison of the objective function values produced by LPT- and RLPT-scheduling presented in Subsection 2.6.3. There, we showed that the RLPT-heuristic is not capable of generating a better schedule than the LPT-heuristic for any of the three objective functions $C_{max}, C_{min}, C_\Delta$. So, an intuitive conjecture might be that (weak) monotonicity exists in all or at least one of the sequences $\left(C_{max}^{k-LPT}\right)_{k=1}^{m}$, $\left(C_{min}^{k-LPT}\right)_{k=1}^{m}$ or $\left(C_\Delta^{k-LPT}\right)_{k=1}^{m}$ of objective function values.

The following $(m, m+1)$-instance I with processing times

$$t_1 = \ldots = t_{m-1} = 2,$$
$$t_m = t_{m+1} = 1,$$

is a counter-example for the previous conjecture. Table 2.3 contains the objective function values of the k-LPT heuristic depending on k.

	C_{max}^{k-LPT}	C_{min}^{k-LPT}	C_Δ^{k-LPT}
$k \mid m$	2	2	0
$k \nmid m$	3	1	2

Table 2.3: Objective function values for I

As $1 \mid m, m \mid m$ and $m \geq 3$, (weak) monotonicity in the corresponding objective function values is not fulfilled for the $(m, m+1)$-instance I.

Chapter 3

Potentially Optimal Makespan Schedules

Despite the NP-hardness of each of the three specific Identical Machine Scheduling Problems (IMSP's) under investigation we dedicate the following two chapters to optimal scheduling and some theoretical aspects.

The present chapter mainly concentrates on the case of two identical machines. As the three objectives introduced in Section 1.3 are equivalent in case of two identical machines, we take the makespan-minimization problem as the representative one. The main questions of interest are:

- How many out of the 2^{n-1} schedules can be optimal, and

- how can they be characterized?

After some introductory words, we answer these two questions in Section 3.2.

Chapter 3 ends with a result concerning convexity which is presented in Section 3.3.

3.1   Introduction

This section is meant to introduce the term *potential optimality* and to formulate the two relevant questions that are to be analyzed here.

At first, we remark that we assume the processing times to be positive and pairwise different throughout this chapter (unless anything else is mentioned), i. e.,

$$t_1 > t_2 > \ldots > t_n > 0 \quad (n \geq 3).$$

So, dummy-jobs are not considered here.

We will see that an extension of our results concerning the case $t_1 > \ldots > t_n > 0$ to the case $t_1 \geq \ldots \geq t_n > 0$ is easily managed.

As no dummy-jobs are considered in this chapter, n denotes the number of non-dummy-jobs. In this case, remember that n needs not to be a multiple of m. The cases $n = 1$ and $n = 2$ are trivial.

As we do not compare different algorithms in this chapter, the upper index on objective function values refers directly to a given schedule.

Definition 3.1.1 (potentially F-optimal)
*Let F be one of the three objective functions C_{max}, C_{min} or C_Δ for the IMSP with m machines and n jobs. A schedule S is called **potentially F-optimal** if n feasible processing times exist so that S is optimal concerning the objective function F.*

Example 3.1.2
Assume $m = 3$ and $n = 6$. Then, the schedule $S = (1, 2, 3, 3, 2, 1)$ is potentially C_{max}-optimal as S is optimal for the $(3, 6)$-instance with processing times $t_j = 7 - j$ $(j = 1, \ldots, 6)$. Obviously, S is also potentially C_{min}-optimal and potentially C_Δ-optimal.

By definition 3.1.1 two interesting questions arise immediately:

1. Depending on m (and n), is it possible to find an elegant characterization of potentially F-optimal schedules for any of the three objectives?

2. How many schedules are potentially F-optimal for fixed m and fixed n?

We precisely deal with these two questions in case of two identical machines in the next section.

3.2 Potentially Optimal Schedules in Case of Two Machines

As already mentioned, the three objectives minimizing C_{max}, maximizing C_{min}, and minimizing C_Δ are equivalent in case of two identical machines. So, instead of saying that a schedule is *potentially F-optimal* we simply say **potentially optimal** if the case $m = 2$ is considered.

The whole subsequent analysis in this chapter is done by representing S as a one-dimensional path P_S. Remember that we are representing schedules as strings of length n over the set $\{1, \ldots, m\}$ throughout this thesis and that we are considering only non-permuted schedules (cf. Section 1.2). This notation has been chosen as it allows a simple transformation of the string-representation to the corresponding path-representation of a schedule. The transformation consists of two steps[1] and works as follows in case $m = 2$. In step one, we transform the $\{1, 2\}$-string S to an $\{1, -1\}$-string S^{trans} in the following way:

$$S^{trans}(j) = \begin{cases} 1, & \text{if } S(j) = 1, \\ -1, & \text{if } S(j) = 2. \end{cases}$$

In step two, we define the corresponding one-dimensional path P_S as

$$P_S(j) = \sum_{k=1}^{j} S^{trans}(k) \quad \text{for} \quad j = 1, \ldots, n$$

and we set $P_S(0) := 0$. So, the path P_S is also represented as a string of length $(n + 1)$.

Note that $P_S(j)$ $(j = 1, \ldots, n)$ is exactly the difference between the number of jobs that are assigned to machine 1 and the number of jobs that are assigned to machine 2 in schedule S after the assignment of the jobs $1, \ldots, j$. Furthermore, remember that we assign the longest job to machine 1 without loss of generality, i. e., $P_S(1) = 1$ for any path.

According to the (graphical) path-representation (cf. Example 3.2.2 on page 74), we use the self-explanatory term *zero-line* or *horizontal-line* in order to locate the path. Moreover, we use the terms *positive* and *negative sector*. So, if $P_S(j) > 0$ for some j we say that the path moves above the zero-line or in the positive sector (at position j). More detailed, if $P_S(j) = k$, we say

---

[1] Of course, it is possible to combine the two steps to one step.

that the path moves at *level* k (at position j). We use the term *crossing* the zero-line at position j if $P_S(j) = 0$ and $((P_S(j-1) = -1$ and $P_S(j+1) = 1)$ or $(P_S(j-1) = 1$ and $P_S(j+1) = -1))$.

Definition 3.2.1
The set $S_{NEG}(n)$ contains all (non-permuted) schedules $S \in \{1,2\}^n$ whose corresponding path P_S moves at least at one position in the negative sector. More formally:

$$S_{NEG}(n) = \Big\{ S \mid P_S(j) < 0 \ \text{for some} \ j \in \{3, \ldots, n\} \Big\}.$$

If no further specification is needed we write S_{NEG} instead of $S_{NEG}(n)$.

Example 3.2.2
Let $n = 6$ and $S = (1, 2, 2, 1, 2, 1)$. Then, $S^{trans} = (1, -1, -1, 1, -1, 1)$ and $P_S = (0, 1, 0, -1, 0, -1, 0)$. The path P_S is illustrated as follows:

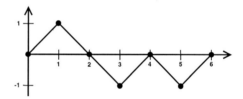

Of course, this schedule is contained in $S_{NEG}(6)$.

The following two theorems give a total characterization of potentially optimal schedules in case of two machines (and pairwise different processing times).

Theorem 3.2.3
Let S be a schedule which is not an element of $S_{NEG}(n)$ $(n \geq 3)$. Then, S is not potentially optimal.

Proof
Let $S \notin S_{NEG}$, i. e., the corresponding path P_S does not move in the negative sector at any position. Let $J_1(S) = \{i_1, \ldots, i_r\}$ denote the set of jobs that are assigned to machine 1 in S. Analogously, $J_2(S) = \{j_1, \ldots, j_s\}$ denotes the set of jobs that are assigned to machine 2 in S. Then, $r \geq s$ and $r + s = n$. We assume $1 = i_1 < i_2 < \ldots < i_r$ and $j_1 < \ldots < j_s$, without loss of

generality. Since $S \notin S_{NEG}$, we can conclude:

$$i_1 < j_1, i_2 < j_2, \ldots, i_s < j_s.$$

Thus, $C_{max}^S = \sum_{k=1}^{r} t_{i_k}$.

In order to prove that S is not potentially optimal, we construct a schedule $\bar{S}$ that is shorter than S for any feasible instance. We distinguish three cases.

Case 1: $s = 0$.

This case is trivial. Here, $J_1(S) = \{1, \ldots, n\}$ and $J_2(S) = \emptyset$. Consider a schedule $\bar{S}$ with $J_1(\bar{S}) = \{1, 3, \ldots, n\}$ and $J_2(\bar{S}) = \{2\}$. Then, it follows immediately $C_{max}^{\bar{S}} < C_{max}^S$.

Case 2: $s = 1$.

Here, assume $J_1(S) = \{1 = i_1, i_2, \ldots, i_{n-1}\}$ and $J_2(S) = \{j\}$. Consider a schedule $\bar{S}$ with $J_1(\bar{S}) = \{i_1, i_3, \ldots, i_{n-1}\}$ and $J_2(\bar{S}) = \{i_2, j\}$. Then, $C_{max}^S = t_1 + \sum_{k=2}^{n-1} t_{i_k}$ and $C_{max}^{\bar{S}} \in \left\{ t_1 + \sum_{k=3}^{n-1} t_{i_k}, t_{i_2} + t_j \right\}$.

It is readily verified that $C_{max}^{\bar{S}} < C_{max}^S$.

Case 3: $s \geq 2$.

Consider a schedule $\bar{S}$ with $J_1(\bar{S}) = \{i_1, \ldots, i_{s-1}, i_{s+1}, \ldots, i_r, j_s\}$ and $J_2(\bar{S}) = \{j_1, \ldots, j_{s-1}, i_s\}$. Then,

$$C_{max}^{\bar{S}} \in \left\{ \sum_{k=1}^{s-1} t_{i_k} + t_{j_s} + \sum_{k=s+1}^{r} t_{i_k}, \sum_{k=1}^{s-1} t_{j_k} + t_{i_s} \right\}.$$

Provided that $C_{max}^{\bar{S}} = \sum_{k=1}^{s-1} t_{i_k} + t_{j_s} + \sum_{k=s+1}^{r} t_{i_k}$, we can conclude:

$$\sum_{k=1}^{s-1} t_{i_k} + t_{j_s} + \sum_{k=s+1}^{r} t_{i_k} \underset{t_{j_s} < t_{i_s}}{<} \sum_{k=1}^{r} t_{i_k} = C_{max}^S.$$

In the other case $C_{max}^{\bar{S}} = \sum_{k=1}^{s-1} t_{j_k} + t_{i_s}$, we can conclude:

$$\sum_{k=1}^{s-1} t_{j_k} + t_{i_s} \underset{t_{j_k} < t_{i_k}}{<} \sum_{k=1}^{s-1} t_{i_k} + t_{i_s} = \sum_{k=1}^{s} t_{i_k} \underset{s \leq r}{\leq} \sum_{k=1}^{r} t_{i_k} = C_{max}^S.$$

Thus, $C_{max}^{\bar{S}} < C_{max}^S$.

This completes the proof of the theorem as in each of the three main cases a shorter schedule $\bar{S}$ exists (for any feasible $(2, n)$-instance). ∎

Corollary 3.2.4
*There exists no $(2,n)$-instance with pairwise different processing times so that
the SPT-LS-rule generates a C_{max}-optimal schedule.*

Proof
The proof is clear. By application of Corollary 2.4.2 (cf. page 27) it follows
directly that the corresponding path of the SPT-LS-schedule does not move
in the negative sector at any position.   ■

Theorem 3.2.5
*Let S be a schedule which is an element of $S_{NEG}(n)$ $(n \geq 3)$. Then, S is
potentially optimal.*

Proof [long; end on page 83]
This theorem is proved by presenting a construction scheme of the processing
times for each schedule $S \in S_{NEG}$ so that $C_1^S = C_2^S$ holds for the job-system
that is constructed, i. e., by equality S is optimal for this job-system and
thus S is potentially optimal.
We distinguish five cases, and we consider a schedule $S \in S_{NEG}(n)$ in the
following of this proof. By j_{neg}^S we denote the maximum (according to the
index) of all jobs j which fulfill $P_S(j) = 0$ and $P_S(j-1) = -1$ (if existent).
More formally:

$$j_{neg}^S = \max\{j \in \{1,\ldots,n\} \mid P_S(j) = 0 \ and \ P_S(j-1) = -1\}.$$

If no such job exists in S, we set $j_{neg}^S := \infty$.
Analogously, we define j_{pos}^S as the maximum of all job-indices which fulfill
$P_S(j) = 0$ and $P_S(j-1) = 1$. More formally:

$$j_{pos}^S = \max\{j \in \{1,\ldots,n\} \mid P_S(j) = 0 \ and \ P_S(j-1) = 1\}.$$

Note that j_{pos}^S exists for any $S \in S_{NEG}$.

Case 1: $j_{neg}^S = n$.

In this case, each of the two machines processes $\frac{n}{2}$ jobs. Moreover, machine
1 processes the longest and the shortest job. Here, the idea is to use these
two jobs to exactly balance the machine completion times of machine 1
and machine 2.
The construction scheme for the t_j's looks as follows. The parameters are
$\varepsilon > 0$ and $x > 0$. The way they should be chosen is explained afterwards.

j	1	2	3	$\ldots$	$n-2$	$n-1$	n
t_j	$1 + \frac{n^2}{2}\varepsilon$	$1 + (n-2)\varepsilon$	$1 + (n-3)\varepsilon$	$\ldots$	$1 + 2\varepsilon$	$1 + \varepsilon$	$1 - x\varepsilon$

The idea of this scheme is the following. As job 1 and job n are processed by the same machine, we choose the coefficient of the ε-term of t_1 large enough and balance the two machine completion times using the parameter x. More precisely, we simply choose $t_1 = 1 + \frac{n^2}{2}\varepsilon$. This ensures that x is positive as $\frac{n^2}{2} > \sum_{j=2}^{\frac{n}{2}+1} (n-j) = \frac{3}{8}n^2 - \frac{3}{4}n$.
As we claim $\varepsilon > 0$, the positivity of x is a sufficient condition for $t_n < t_{n-1}$. Then, we choose

$$x = \frac{n^2}{2} + \sum_{j \in \{2,\ldots,n-1\}:S(j)=1} (n-j) - \sum_{j \in \{2,\ldots,n-1\}:S(j)=2} (n-j),$$

which is greater than zero.
As t_n must be positive, we choose ε so that $1 - x\varepsilon > 0$, i. e., $\varepsilon \in (0, \frac{1}{x})$. This yields $C_1^S = C_2^S$.

Example 3.2.6
Let $n = 6$ and $S = (1, 2, 2, 1, 2, 1)$ (cf. Example 3.2.2 on page 74).

j	*1*	*2*	*3*	*4*	*5*	*6*
t_j	*$1+18\varepsilon$*	*$1+4\varepsilon$*	*$1+3\varepsilon$*	*$1+2\varepsilon$*	*$1+\varepsilon$*	*$1-x\varepsilon$*
$\tilde{t}_j$	*33*	*19*	*18*	*17*	*16*	*3*

Here, we get $x = 18 + 2 - (4 + 3 + 1) = 12$ and ε can be chosen from the interval $(0, \frac{1}{12})$. So, choosing $\varepsilon = \frac{1}{15}$ and multiplying each t_j by $\frac{1}{\varepsilon}$ leads to the processing times $\tilde{t}_j$. Regarding the $\tilde{t}_j$'s we get $C_1^S = 53 = C_2^S$. Thus, S is potentially optimal as it is optimal for the given set of $\tilde{t}_j$-values. Of course, S is also optimal concerning the t_j's since multiplication of the processing times by a positive constant has no influence on the optimality of a schedule.

Case 2: $j_{neg}^S < n$ and $P_S(j) \geq 0$ for all $j \in \{j_{neg}^S + 1, \ldots, n\}$.

In this case, we know that at least as many jobs out of $\{j_{neg}^S + 1, \ldots, n\}$ are processed by machine 1 as by machine 2. Moreover, considering the longest j_{neg}^S jobs, we can apply the scheme of Case 1 to construct processing times so that $C_1^S(j_{neg}^S) = C_2^S(j_{neg}^S)$.
We use this fact and adapt the scheme slightly by modifying the processing time of job 1 for balancing reasons. The construction scheme has four

positive parameters $\varepsilon, \delta, x, y$. The scheme is as follows:

j	1	2	$\ldots$	$j_{neg}^S - 1$	j_{neg}^S
t_j	$1 + \frac{(j_{neg}^S)^2}{2}\varepsilon - y\delta$	$1 + (j_{neg}^S - 2)\varepsilon$	$\ldots$	$1 + \varepsilon$	$1 - x\varepsilon$

j	$j_{neg}^S + 1$	$\ldots$	n
t_j	$(n - j_{neg}^S)\delta$	$\ldots$	δ

At first, the value of x is chosen as in Case 1 so that $C_1^S(j_{neg}^S) = C_2^S(j_{neg}^S)$, i. e.,

$$x = \frac{(j_{neg}^S)^2}{2} + \sum_{j \in \{2,\ldots,j_{neg}^S - 1\}: S(j)=1} (j_{neg}^S - j) - \sum_{j \in \{2,\ldots,j_{neg}^S - 1\}: S(j)=2} (j_{neg}^S - j).$$

Then, we choose $\varepsilon \in (0, \frac{1}{x})$. In a next step, we determine y as

$$y = \sum_{j \in \{j_{neg}^S + 1,\ldots,n\}: S(j)=1} (n - j + 1) - \sum_{j \in \{j_{neg}^S + 1,\ldots,n\}: S(j)=2} (n - j + 1).$$

As $P_S(j) \geq 0$ for $j = j_{neg}^S + 1, \ldots, n$ and the processing times are pairwise different, we know that $y > 0$. Now, we choose $\delta > 0$ so that the following two conditions are fulfilled:

(i) $t_{j_{neg}^S} > t_{j_{neg}^S + 1} \Leftrightarrow 1 - x\varepsilon > (n - j_{neg}^S)\delta \Leftrightarrow \delta < \frac{1 - x\varepsilon}{n - j_{neg}^S}$

(ii) $t_1 > t_2 \Leftrightarrow 1 + \frac{(j_{neg}^S)^2}{2}\varepsilon - y\delta > 1 + (j_{neg}^S - 2)\varepsilon \Leftrightarrow \delta < \left(\frac{(j_{neg}^S)^2}{2} - j_{neg}^S + 2 \right) \frac{\varepsilon}{y}.$

By choosing $\delta \in \left(0, \min \left\{ \frac{1 - x\varepsilon}{n - j_{neg}^S}, \left(\frac{(j_{neg}^S)^2}{2} - j_{neg}^S + 2 \right) \frac{\varepsilon}{y} \right\} \right)$ we have constructed feasible processing times so that $C_1^S = C_2^S$.

Example 3.2.7
Let $n = 8$ and $S = (1, 2, 2, 1, 2, 1, 1, 1)$. Here, we have $j_{neg}^S = 6$ and application of the construction scheme yields

j	1	2	3	4	5	6	7	8
t_j	$1 + 18\varepsilon - y\delta$	$1 + 4\varepsilon$	$1 + 3\varepsilon$	$1 + 2\varepsilon$	$1 + \varepsilon$	$1 - x\varepsilon$	2δ	δ
$\tilde{t}_j$	615	380	360	340	320	60	30	15

As the longest six jobs are assigned in the same way as in Example 3.2.6 of Case 1, we know that $x = 12$ and $\varepsilon \in (0, \frac{1}{12})$. Again, we choose $\varepsilon = \frac{1}{15}$. Calculating y leads to $y = 3$. Thus, choosing $\delta \in (0, \min\{\frac{1}{10}, \frac{14}{45}\})$ leads to feasible processing times so that $C_1^S = C_2^S$. For instance, we choose $\delta = \frac{1}{20}$. Multiplying the t_j's by $\frac{1}{\varepsilon\delta}$ leads to the processing times $\tilde{t}_j$'s for which we get $C_1^S = 1060 = C_2^S$.

Case 3: $j_{neg}^S < n$ and $P_S(j) < 0$ for all $j \in \{j_{neg}^S + 1, \ldots, n\}$.

In this case, we know that in the partial schedule S machine 2 processes more jobs than machine 1 after the assignment of j jobs ($j \in \{j_{neg}^S + 1, \ldots, n\}$).

Considering the longest j_{neg}^S jobs, we can apply the scheme of Case 1 to construct processing times so that $C_1^S(j_{neg}^S) = C_2^S(j_{neg}^S)$.

Again, we use this fact and adapt the scheme slightly by modifying the processing time of job 1 for balancing reasons. The construction scheme is the same as the one in Case 2 except for job 1. This time, t_1 will be enlarged. Again, we use four positive parameters $\varepsilon, \delta, x, y$. Then, the scheme is as follows:

j	1	2	$\cdots$	$j_{neg}^S - 1$	j_{neg}^S
t_j	$1 + \frac{(j_{neg}^S)^2}{2}\varepsilon + y\delta$	$1 + (j_{neg}^S - 2)\varepsilon$	$\cdots$	$1 + \varepsilon$	$1 - x\varepsilon$

j	$j_{neg}^S + 1$	$\cdots$	n
t_j	$(n - j_{neg}^S)\delta$	$\cdots$	δ

The parameter x and the feasible values for ε are determined as in Case 2. This time, y is determined as

$$y = \sum_{j \in \{j_{neg}^S+1,\ldots,n\}:S(j)=2} (n - j + 1) - \sum_{j \in \{j_{neg}^S+1,\ldots,n\}:S(j)=1} (n - j + 1),$$

which yields $y > 0$. The parameter δ can be chosen from the interval $(0, \frac{1 - x\varepsilon}{n - j_{neg}^S})$. This ensures $t_{j_{neg}^S} > t_{j_{neg}^S + 1}$.

Example 3.2.8
Let $n = 8$ and $S = (1, 2, 2, 1, 2, 1, 2, 2)$. Here, we have $j_{neg}^S = 6$ and application of the construction scheme yields

j	1	2	3	4	5	6	7	8
t_j	$1+18\varepsilon+y\delta$	$1+4\varepsilon$	$1+3\varepsilon$	$1+2\varepsilon$	$1+\varepsilon$	1-$x\varepsilon$	2δ	δ
$\tilde{t}_j$	705	380	360	340	320	60	30	15

Analogously to the previous cases, we get $x = 12$ and we can choose $\varepsilon \in (0, \frac{1}{12})$. Again, we choose $\varepsilon = \frac{1}{15}$. Calculating y leads to $y = 3$. Thus, choosing $\delta \in (0, \frac{1}{10})$ leads to feasible processing times so that $C_1^S = C_2^S$. Again, we choose $\delta = \frac{1}{20}$. Multiplying the t_j's by $\frac{1}{\varepsilon\delta}$ leads to the processing times $\tilde{t}_j$'s for which we get $C_1^S = 1105 = C_2^S$.

Case 4: $j^S_{neg} < j^S_{pos} < n$ and $P_S(j) < 0$ for $j \in \{j^S_{pos} + 1, \ldots, n\}$.

In this case, the corresponding path crosses the zero-line at least once coming from the negative sector, and before crossing the zero-line the last time the path moved in the positive sector. At all positions $j > j^S_{pos}$, the path moves in the negative sector.

So, considering the longest j^S_{pos} jobs, we can apply the scheme of Case 2 to construct processing times so that $C^S_1(j^S_{pos}) = C^S_2(j^S_{pos})$. We use this fact and adapt this scheme slightly by modifying the processing time of job 1 for balancing reasons. Now, the scheme has six positive parameters $\varepsilon, \delta, \gamma, x, y, z$. The scheme looks as follows:

j	1	2	$\cdots$	$j^S_{neg} - 1$	j^S_{neg}
t_j	$1 + \frac{(j^S_{neg})^2}{2}\varepsilon - y\delta + z\gamma$	$1 + (j^S_{neg} - 2)\varepsilon$	$\cdots$	$1 + \varepsilon$	$1 - x\varepsilon$

j	$j^S_{neg} + 1$	$\cdots$	j^S_{pos}
t_j	$(j^S_{pos} - j^S_{neg})\delta$	$\cdots$	δ

j	$j^S_{pos} + 1$	$\cdots$	n
t_j	$(n - j^S_{pos})\gamma$	$\cdots$	γ

The parameter x is determined as in Case 2, and y is also determined analogously to Case 2, i. e.,

$$y = \sum_{j \in \{j^S_{neg}+1, \ldots, j^S_{pos}\}: S(j)=1} (j^S_{pos} - j + 1) - \sum_{j \in \{j^S_{neg}+1, \ldots, j^S_{pos}\}: S(j)=2} (j^S_{pos} - j + 1).$$

The parameter ε can be chosen from $(0, \frac{1}{x})$ and δ can be chosen from $\left(0, \min\left\{\frac{1 - x\varepsilon}{j^S_{pos} - j^S_{neg}}, \left(\frac{(j^S_{neg})^2}{2} - j^S_{neg} + 2\right)\frac{\varepsilon}{y}\right\}\right)$. This yields $C^S_1(j^S_{pos}) = C^S_2(j^S_{pos})$. Then, we determine z as

$$z = \sum_{j \in \{j^S_{pos}+1, \ldots, n\}: S(j)=2} (n - j + 1) - \sum_{j \in \{j^S_{pos}+1, \ldots, n\}: S(j)=1} (n - j + 1),$$

which yields $z > 0$. The parameter γ can be chosen from $(0, \frac{\delta}{n - j^S_{pos}})$, which ensures $t_{j^S_{pos}} > t_{j^S_{pos}+1}$.

Example 3.2.9
Let $n = 10$ and $S = (1, 2, 2, 1, 2, 1, 1, 2, 2, 2)$. Here, we have $6 = j^S_{neg} < j^S_{pos} = 8$. Application of the construction scheme yields

j	1	2	3	4	5	6
t_j	$1+18\varepsilon - y\delta + z\gamma$	$1+4\varepsilon$	$1+3\varepsilon$	$1+2\varepsilon$	$1+\varepsilon$	$1-x\varepsilon$
$\tilde{t}_j$	33150	19000	18000	17000	16000	3000

j	7	8	9	10
t_j	2δ	δ	2γ	γ
$\tilde{t}_j$	1500	750	600	300

Analogously to the previous cases, we get $x = 12$ and we can choose $\varepsilon \in (0, \frac{1}{12})$. Again, we choose $\varepsilon = \frac{1}{15}$. Calculating y leads to $y = 1$. Thus, choosing $\delta \in (0, \min\{\frac{1}{10}, \frac{14}{15}\})$ leads to feasible processing times so that $C_1^S(8) = C_2^S(8)$. Again, we choose $\delta = \frac{1}{20}$. The calculation of z leads to $z = 3$. Then, γ can be chosen from $(0, \frac{1}{40})$. For instance, we choose $\gamma = \frac{1}{50}$. Multiplying the t_j's by $\frac{1}{\varepsilon\delta\gamma}$ leads to the processing times $\tilde{t}_j$'s for which we get $C_1^S = 54650 = C_2^S$.

Case 5: $j_{neg}^S = \infty$.

In this case, the corresponding path crosses the zero-line exactly once and moves only in the negative sector afterwards. We distinguish the two subcases $S(n) = 1$ and $S(n) = 2$.

We start with $S(n) = 1$. Here, we use a construction scheme which is similar to the one of Case 1. The difference is that the processing time of job 1 is increased for balancing reasons. The scheme has the two positive parameters ε and x:

j	1	2	3	$\ldots$	$n-2$
t_j	$1 + \frac{n^2}{2}\varepsilon + \lvert P_S(n) \rvert$	$1 + (n-2)\varepsilon$	$1 + (n-3)\varepsilon$	$\ldots$	$1 + 2\varepsilon$

j	$n-1$	n
t_j	$1 + \varepsilon$	$1 - x\varepsilon$

The parameter x is determined as in Case 1, and x is positive since $\frac{n^2}{2} > \sum_{j=2}^{n-1}(n-j)$ for all $n \geq 3$. The parameter ε can be chosen from $(0, \frac{1}{x})$. The summand $\lvert P_S(n) \rvert$ in t_1 is due to the fact that machine 2 processes exactly $P_S(n)$ jobs more than machine 1. So, we get $C_1^S = C_2^S$.

Example 3.2.10
Let $n = 8$ and $S = (1, 2, 2, 2, 1, 2, 2, 1)$. Here, we have $j_{neg}^S = \infty$ and $\lvert P_S(n) \rvert = 2$. Application of the construction scheme yields

j	1	2	3	4	5	6
t_j	$1+32\varepsilon+\lfloor P_S(n)\rfloor$	$1+6\varepsilon$	$1+5\varepsilon$	$1+4\varepsilon$	$1+3\varepsilon$	$1+2\varepsilon$
$\tilde t_j$	92	26	25	24	23	22

j	7	8
t_j	$1+\varepsilon$	$1\text{-}x\varepsilon$
$\tilde t_j$	21	3

Here, we get $x = 17$ and ε can be chosen from $(0, \frac{1}{17})$. We choose $\varepsilon = \frac{1}{20}$. Multiplying the t_j's by $\frac{1}{\varepsilon}$ leads to the processing times $\tilde t_j$'s for which we get $C_1^S = 118 = C_2^S$.

In the other subcase, i. e., $S(n) = 2$, we simply add a fictive job $(n+1)$ to the construction scheme. We assume this job to be processed by machine 1, and we modify the construction scheme of the previous subcase $S(n) = 1$. Again, the scheme has two positive parameters ε and x. Including the fictive job, it looks as follows (with f_j denoting the temporary or fictive processing times):

j	1	2	3	$\ldots$	$n-1$
f_j	$1+\frac{n^2}{2}\varepsilon+(\lfloor P_S(n)\rfloor-1)$	$1+(n-1)\varepsilon$	$1+(n-2)\varepsilon$	$\ldots$	$1+2\varepsilon$

j	n	$n+1$
f_j	$1+\varepsilon$	$1-x\varepsilon$

The parameter x is determined as

$$x = \frac{n^2}{2} + \sum_{j\in\{2,\ldots,n\}:S(j)=1} (n-j+1) - \sum_{j\in\{2,\ldots,n\}:S(j)=2} (n-j+1).$$

Since $\frac{n^2}{2} > \sum_{j=2}^{n}(n-j+1)$ for all $n \geq 3$, x is positive. The parameter ε can be chosen analogously to the previous subcase. The summand $(\lfloor P_S(n)\rfloor - 1)$ in t_1 leads to equal machine completion times concerning the fictive system.

In order to construct processing times t_j so that $C_1^S = C_2^S$ holds for the primal system, we add the processing time f_{n+1} of the fictive job to f_1, i. e., $t_1 = f_1 + f_{n+1}$. All other processing times remain unchanged, i. e., $t_j = f_j$ for $j = 2, \ldots, n$.

Example 3.2.11
Let $n = 6$ and $S = (1, 2, 2, 2, 1, 2)$. Here, we have $j_{neg}^S = \infty$. Application of the construction scheme yields

j	1	2	3	4	5
f_j	$1+18\varepsilon+(\lceil P_S(n)\rceil-1)$	$1+5\varepsilon$	$1+4\varepsilon$	$1+3\varepsilon$	$1+2\varepsilon$
t_j	$3+11\varepsilon$	$1+5\varepsilon$	$1+4\varepsilon$	$1+3\varepsilon$	$1+2\varepsilon$
$\tilde{t}_j$	41	15	14	13	12

j	6	7
f_j	$1+\varepsilon$	$1-x\varepsilon$
t_j	$1+\varepsilon$	
$\tilde{t}_j$	11	

Now, we get $x = 7$ and ε can be chosen from $(0, \frac{1}{7})$. We choose $\varepsilon = \frac{1}{10}$. Multiplying the t_j's by $\frac{1}{\varepsilon}$ leads to the processing times $\tilde{t}_j$'s for which we get $C_1^S = 53 = C_2^S$.

This completes the proof of Theorem 3.2.5.   ∎

We remark that the construction schemes presented in the previous proof do in general not generate processing times so that a schedule is uniquely optimal for them. Consider the schedule $S_1 = (1, 2, 1, 2, 2, 1, 2, 1)$ for which the scheme of Case 1 can be applied in order to construct feasible processing times so that $C_1^{S_1} = C_2^{S_1}$. Then, it is readily verified that for the same processing times the schedule $S_2 = (1, 2, 2, 1, 1, 2, 2, 1)$ is optimal, too. Nevertheless, in case of two machines we conjecture that every potentially optimal schedule is also potentially unique optimal, i. e., feasible processing times exist so that a schedule is uniquely optimal.

By the Theorems 3.2.3 (cf. page 74) and 3.2.5 (cf. page 76), we know that a schedule S is potentially optimal if and only if S is an element of S_{NEG}. Let $\overline{S_{NEG}}$ denote the set of schedules that are not potentially optimal. We know that $|S_{NEG}(n)| + |\overline{S_{NEG}(n)}| = 2^{n-1}$. Instead of a direct determination of the number of potentially optimal schedules, we determine the cardinality of the set $\overline{S_{NEG}}$. The arising recursion can easily be solved by application of the Catalan numbers [Ste02, pages 169-172]. A helpful interpretation of the Catalan numbers is the following:
Concerning a $(2, 2k)$-instance ($k \in \mathbb{N}$), the number of schedules whose corresponding path[2] ends on the zero-line at position $2k$ and does not move below the zero-line at any intermediate position is given by the k-th Catalan number $Cata_k = \frac{1}{k+1}\binom{2k}{k}$.

---

[2]Remember that all corresponding paths start on the zero-line at position 0.

Theorem 3.2.12

Let $n \geq 3$. Then, the number of (non-permuted) schedules that are not potentially optimal is

$$|\overline{S_{NEG}(n)}| = \binom{n}{\lceil \frac{n}{2} \rceil}.$$

Proof

The proof works by induction in n.

Base of Induction: $n = 3$

The set $\overline{S_{NEG}(3)}$ consists of the following three schedules:

$$\overline{S_{NEG}(3)} = \{(1,1,1), (1,1,2), (1,2,1)\}.$$

Moreover, $\binom{3}{\lceil \frac{3}{2} \rceil} = 3$.

Step of Induction: $n \rightarrow n + 1$

Depending on the parity of $(n + 1)$ we consider the following two cases.

Case 1: $2 \mid (n + 1)$.

In this case, all paths in $\overline{S_{NEG}(n)}$ end at an odd positive level. Then, each of these paths can be continued in each of the two possible ways to receive a path that is contained in $\overline{S_{NEG}(n + 1)}$. So,

$$|\overline{S_{NEG}(n+1)}| \overset{Ind.}{\underset{Hypo.}{=}} 2|\overline{S_{NEG}(n)}| \overset{Ind.}{\underset{Hypo.}{=}} 2\binom{n}{\lceil \frac{n}{2} \rceil} = \binom{n}{\frac{n+1}{2}} + \binom{n}{\frac{n+1}{2}} =$$

$$= \binom{n}{\frac{n+1}{2}} + \binom{n}{\frac{n-1}{2}} = \binom{n}{\frac{n+1}{2}} + \binom{n}{\frac{n+1}{2} - 1} = \binom{n+1}{\frac{n+1}{2}} =$$

$$= \binom{n+1}{\lceil \frac{n+1}{2} \rceil}.$$

Case 2: $2 \nmid (n + 1)$.

In this case, all paths in $\overline{S_{NEG}(n)}$ end at an even positive level or at level 0. Then, each of the paths that ends at an even positive level can be continued in each of the two possible ways to receive a path that is contained in $\overline{S_{NEG}(n + 1)}$. In contrast, each of the paths that ends at level 0 can be continued in only one possible way. Remember that the number of these paths is given by $Cata_{\frac{n}{2}}$. So,

$$\overline{|S_{NEG}(n+1)|} = 2\overline{|S_{NEG}(n)|} - Cata_{\frac{n}{2}} \overset{Ind.}{\underset{Hypo.}{=}} 2\binom{n}{\lceil\frac{n}{2}\rceil} - \frac{1}{\frac{n}{2}+1}\binom{n}{\frac{n}{2}} =$$

$$= \left(2 - \frac{1}{\frac{n}{2}+1}\right)\binom{n}{\frac{n}{2}} = \frac{n+1}{\frac{n}{2}+1}\binom{n}{\frac{n}{2}} = \frac{(n+1)!}{(\frac{n}{2})!(\frac{n}{2}+1)!} =$$

$$= \binom{n+1}{\frac{n}{2}+1} = \binom{n+1}{\lceil\frac{n+1}{2}\rceil}.$$

∎

The application of Stirling's approximation [Wal01, page 352], i. e., $n! \approx \sqrt{2\pi n}(\frac{n}{e})^n$, yields that the rate of not potentially optimal schedules is approximately $\frac{2\sqrt{2}}{\sqrt{\pi}}\frac{1}{\sqrt{n}}$ for large n.

Extension

The extension concerns the processing times. So far, we assumed that the processing times are pairwise different. Now, we briefly extend our results on potentially optimal schedules to the case that the processing times are not pairwise different, i. e., $t_1 \geq t_2 \geq \ldots \geq t_n > 0$. We still consider two machines. According to the Theorems 3.2.3 and 3.2.5 the following is readily verified:
If n is even, then each schedule whose corresponding path moves only in the non-negative sector and ends on the zero-line is also potentially optimal.
If n is odd, then each schedule whose corresponding path moves only in the non-negative sector and ends on level 1 is also potentially optimal. Feasible processing times are $t_j = t > 0$ for all jobs $j = 1, \ldots, n$.
So, the number of potentially optimal schedules increases by $Cata_{\frac{n}{2}}$ in case n is even and by $Cata_{\frac{n+1}{2}}$ in case n is odd.

3.3   A Result Concerning Convexity

We want to finish this chapter by taking a brief look at the following issue. Assume a schedule $S \in \{1,2\}^n$ is optimal for two different $(2,n)$-instances I_1 and I_2 (each represented by a vector of processing times). Then, the question is whether S is also optimal for any convex combined instance $I_\lambda := \lambda I_1 + (1-\lambda)I_2$ $(0 < \lambda < 1)$ of I_1 and I_2?

To motivate this question, assume $1 \geq t_1 \geq t_2 \geq \ldots \geq t_n \geq 0$ without loss of generality. Then, we are interested in the question whether (potentially optimal) $(2, n)$-schedules are optimal on convex sets in the polytope $1 \geq t_1 \geq t_2 \geq \ldots \geq t_n \geq 0$.
Furthermore, it would be interesting to know the rate of each (potentially optimal) schedule on the total volume $\frac{1}{n!}$ of the polytope.

Even in case of two machines the question of convexity is to be denied. We show this by an illustrative example (using integral processing times). Consider the instances $I_1 = (22, 16, 10, 8, 6, 6)$ and $I_2 = (22, 10, 10, 10, 2, 2)^3$. Then, $S_1 = (1, 2, 2, 2, 1, 1)$ is (uniquely) optimal concerning instance I_1 as $C_1^{S_1}(I_1) = 34 = C_2^{S_1}(I_1)$, and S_1 is also (uniquely) optimal concerning instance I_2. The machine completion times of schedule S_1 concerning I_2 are $C_1^{S_1}(I_2) = 26$ and $C_2^{S_1}(I_2) = 30$. Any other (non-permuted) schedule results in a makespan of at least 32 concerning I_2. Choosing $\lambda = \frac{1}{2}$ yields the convex combined instance $I_{\frac{1}{2}} = (22, 13, 10, 9, 4, 4)$ for which the machine completion times of S_1 are $C_1^{S_1}(I_{\frac{1}{2}}) = 30$ and $C_2^{S_1}(I_{\frac{1}{2}}) = 32$. This is not optimal as the schedule $S_2 = (1, 2, 2, 1, 2, 2)$ yields a perfect partition of the jobs, i. e., $C_1^{S_2}(I_{\frac{1}{2}}) = 31$ and $C_2^{S_2}(I_{\frac{1}{2}}) = 31$. It is readily verified that S_2 is neither optimal concerning I_1 nor concerning I_2. Furthermore, note that S_1 is a cardinality-balanced schedule, whereas S_2 does not fulfill this property.
Of course, it is possible to modify I_1 and I_2 to $\tilde{I}_1$ and $\tilde{I}_2$ so that the sum of processing times in $\tilde{I}_1$ equals the sum in $\tilde{I}_2$. In our example, one can choose $\tilde{I}_1 = 14 I_1$ and $\tilde{I}_2 = 17 I_2$ which does not influence the optimality of S_1 concerning these instances. Now, choosing $\lambda = \frac{17}{31}$ ensures that S_2 is optimal concerning $\tilde{I}_{\frac{17}{31}}$ as equality

$$\tilde{I}_{\frac{17}{31}} = \frac{17}{31}\tilde{I}_1 + \frac{14}{31}\tilde{I}_2 = \frac{14 \cdot 17}{31}(I_1 + I_2) = \frac{17 \cdot 14 \cdot 2}{31}(\frac{1}{2}I_1 + \frac{1}{2}I_2)$$

holds.
Furthermore, it is possible to construct instances for any fixed number m of machines ($m \geq 3$) based on I_1 and I_2 so that the convexity-property, as used in our sense, is not fulfilled.

In contrast to optimal scheduling, LS-schedules fulfill the convexity-property. More detailed, let L be a priority list of the LS-algorithm for (m, n)-instances and assume a fixed tie-break rule to be applied, e. g., always choosing the machine with smallest index[4].

---

[3]We did not find $(2, n)$-instances with less than six jobs concerning this issue.

[4]Remember (cf. page 15) that this does generally not produce non-permuted schedules. However, this does not matter here.

So, let I_1 and I_2 be two different (m,n)-instances for which L generates the same schedule $S^L = (S^L(1), \dots, S^L(n))$ (by application of a fixed tie-break rule). Assume that $0 \le (k-1) \le n-1$ jobs have already been assigned according to L, and let the next job on the list be job j, i. e., $L(k) = j$. Furthermore, assume that job j is assigned to machine i_j, i. e., $S^L(j) = i_j$. Due to this assignment, the following inequality must be fulfilled for all machines $i \in \{1, \dots, m\} \setminus \{i_j\}$:

$$\sum_{h \in \{1,\dots,k-1\}: S^L(L(h))=i_j} t^I_{L(h)} \le \sum_{h \in \{1,\dots,k-1\}: S^L(L(h))=i} t^I_{L(h)} \quad (I \in \{I_1, I_2\}).$$

So, for each of the n assignments according to L we can formulate $(m-1)$ linear inequalities which are fulfilled concerning I_1 as well as I_2. It is readily verified that they are also fulfilled for any convex-combined instance $I_\lambda = \lambda I_1 + (1-\lambda)I_2$ $(0 < \lambda < 1)$.

Regarding the LS-algorithm and a fixed tie-break rule we can summarize the following:
Let L be a priority list that generates the same schedule S^L for two different instances I_1 and I_2, then L also generates S^L for any convex combined instance I_λ of I_1 and I_2. As mentioned before, this is in contrast to optimal scheduling where the convexity-property is not fulfilled for all $m \ge 2$.

Chapter 4

On Differences Between $C_{max}, C_{min}, C_\Delta$

This short chapter presents smallest and illustrative instances emphasizing the non-equivalence of the three objectives

- minimizing C_{max},
- maximizing C_{min},
- minimizing C_Δ

in case of at least three machines. Thereby, we compare each pair of objective functions, and we do a joint comparison of all three objective functions.

In case of three identical machines, Chapter 4 also contains some experimental results on the probability that the C_Δ-optimal schedule is not C_{max}- or C_{min}-optimal. In such cases, we also determined the differences in the objective function values.

4.1   Introduction

We start with some introductory remarks. Analogously to the previous chapter, we do not consider dummy-jobs here, i. e., n denotes the number of non-dummy-jobs and n needs not to be a multiple of m which is greater than two in this chapter. We mainly concentrate on the case $m = 3$ as this already reveals the non-equivalence of the three objectives considered in this thesis. Moreover, as we are interested in small or more precisely in smallest instances to contribute to the non-equivalence, we shall also clarify in which sense the term *smallest instance* is used here. We will see that this term is connected with a lexicographic order of three properties.

The term smallest instance is related to an (m, n)-instance I where all processing times are integral, i. e., $t_j \in \mathbb{N}$ for $j = 1, \ldots, n$, and

(i) m is as small as possible,

(ii) n is as small as possible and

(iii) t_1 is as small as possible $(t_1 \geq t_2 \geq \ldots \geq t_n > 0)$

so that neither of the optimal schedules concerning an objective function $F \in \{C_{max}, C_{min}, C_{\Delta}\}$ is also optimal concerning another objective function $\bar{F} \in \{C_{max}, C_{min}, C_{\Delta}\} \setminus \{F\}$. Thereby, the properties $(i) - (iii)$ have decreasing weights. So, we can speak more precisely of lexicographic smallest instances.

We also mention that the smallest instances presented in the following sections were found by total enumeration with the help of a computer. This means that we have generated all feasible instances for fixed m, n and t_1. These are $\binom{t_1 - 1}{n - 1}$ different instances in case of pairwise different processing times and $\binom{t_1 + (n-1) - 1}{n - 1}$ different instances in case of not pairwise different processing times.

4.2 C_{max} Versus C_{Δ}

Considering the objective functions C_{max} and C_{Δ}, we can state that a C_{max}-optimal schedule needs not to be C_{Δ}-optimal. This can easily be verified by considering an (m, m)-instance with $p_1 > \sum_{j=2}^{m} p_j$.

Next, we take a look at C_{Δ}-optimal schedules and their performance concerning the objective function C_{max}. Considering any feasible (m, n)-instance I

with $m = 3$ machines and $n \leq 5$ jobs, we have shown that a C_{Δ}-optimal schedule exists which is also C_{max}-optimal for I (proof is omitted in this thesis). Moreover, we suppose that every C_{Δ}-optimal schedule is C_{max}-optimal in case $m = 3$ and $n \leq 5$. This would imply that no C_{Δ}-optimal schedules with different makespans exist in case $m = 3$ and $n \leq 5$.

However, in case of $n = 6$ jobs and $m = 3$ machines the previous statement is not true. The smallest instance with pairwise different processing times so that none of the C_{Δ}-optimal schedules is optimal concerning C_{max} is $I_{C_{max},C_{\Delta}} = (16, 11, 10, 8, 7, 6)$. The next table contains relevant information on the C_{max}-optimal schedule $S^*_{C_{max}}$ and the C_{Δ}-optimal schedule $S^*_{C_{\Delta}}$ which are both unique for this instance (optimality to be seen by hand calculation).

	C_1	C_2	C_3	C_{max}	C_{Δ}
$S^*_{C_{max}} = (1, 2, 2, 3, 3, 3)$	16	21	21	21	5
$S^*_{C_{\Delta}} = (1, 2, 3, 3, 2, 1)$	22	18	18	22	4

Table 4.1: Machine completion times and objective function values of the optimal schedules concerning $I_{C_{max},C_{\Delta}} = (16, 11, 10, 8, 7, 6)$

Regarding the instance $I_{C_{max},C_{\Delta}}$, note that the schedule $S^*_{C_{\Delta}}$ is also optimal concerning the objective function C_{min} .

In case of not pairwise different processing times the smallest instance that we have found is $\tilde{I}_{C_{max},C_{\Delta}} = (13, 9, 9, 6, 6, 6)$. Regarding this instance, the C_{Δ}-optimal schedule is not unique, and the C_{max}-optimal schedule is the same as the one for the instance $I_{C_{max},C_{\Delta}}$. The corresponding machine completion times (C_1, C_2, C_3) are

- $(13, 18, 18)$ of the C_{max}-optimal schedule and

- $(19, 15, 15)$ of a C_{Δ}-optimal schedule.

With the basic example $I_{C_{max},C_{\Delta}}$ the construction of instances with $n = m+3$ jobs and $m \geq 4$ machines so that none of the C_{Δ}-optimal schedules is optimal for C_{max} is quite obvious, for instance by adding jobs (one for each additional machine) with lengths out of the interval $(18, 21)$ to $I_{C_{max},C_{\Delta}}$.

4.3 C_{min} Versus C_Δ

Taking a look at the objective functions C_{min} and C_Δ, we can state that a C_{min}-optimal schedule needs not to be C_Δ-optimal. This can trivially be achieved in job-systems with less than m jobs as any schedule is C_{min}-optimal in such cases but not necessarily C_Δ-optimal.

So, we look at C_Δ-optimal schedules and their performance concerning the objective function C_{min}. In case $m = 3$ and $n \le 5$ we did not find an instance so that a C_Δ-optimal schedule is not C_{min}-optimal (observe that this is a weaker formulation than in Section 4.2).

The smallest instance with pairwise different processing times that we have found so that none of the C_Δ-optimal schedules is also C_{min}-optimal is the $(3, 6)$-instance $I_{C_{min}, C_\Delta} = (17, 11, 10, 8, 7, 6)$. Again, relevant information on the C_{min}-optimal schedule $S^*_{C_{min}}$ and the C_Δ-optimal schedule $S^*_{C_\Delta}$ for this instance are to be found in the following table (optimality to be seen by hand calculation).

	C_1	C_2	C_3	C_{min}	C_Δ
$S^*_{C_{min}} = (1, 2, 3, 3, 2, 1)$	23	18	18	18	5
$S^*_{C_\Delta} = (1, 2, 2, 3, 3, 3)$	17	21	21	17	4

Table 4.2: Machine completion times and objective function values of the optimal schedules concerning $I_{C_{min}, C_\Delta} = (17, 11, 10, 8, 7, 6)$

Regarding the instance I_{C_{min}, C_Δ}, note that the schedule $S^*_{C_\Delta}$ is also optimal concerning the objective function C_{max}.

In case of not pairwise different processing times the smallest instance that we have found is $\tilde{I}_{C_{min}, C_\Delta} = (14, 9, 9, 6, 6, 6)$. Regarding this instance, the C_{min}-optimal schedule is not unique, and the C_Δ-optimal schedule is the same as the one for the instance I_{C_{min}, C_Δ}. The corresponding machine completion times (C_1, C_2, C_3) are

- $(20, 15, 15)$ of a C_{min}-optimal schedule and

- $(14, 18, 18)$ of the C_Δ-optimal schedule.

It is also interesting to note that the instances $I_{C_{max}, C_\Delta} = (16, 11, 10, 8, 7, 6)$ and $I_{C_{min}, C_\Delta} = (17, 11, 10, 8, 7, 6)$ differ only in the processing time of the longest job and that the difference is 1. The same is true for the instances $\tilde{I}_{C_{max}, C_\Delta} = (13, 9, 9, 6, 6, 6)$ and $\tilde{I}_{C_{min}, C_\Delta} = (14, 9, 9, 6, 6, 6)$.

Again, with the basic example I_{C_{min},C_Δ} the construction of instances with $n = m+3$ jobs and $m \geq 4$ machines so that none of the C_Δ-optimal schedules is also C_{min}-optimal is easily managed.

4.4 C_{max} Versus C_{min}

By the two instances I_{C_{max},C_Δ} and I_{C_{min},C_Δ} of the previous two sections, we have already shown that the optimization of the two objective functions C_{max} and C_{min} is not equivalent even in case of three machines.

In this context, the smallest job-system with pairwise different processing times that we have found is the $(3,6)$-instance $I_{C_{max},C_{min}} = (12, 8, 7, 6, 5, 4)$. Again, relevant information on the C_{max}-optimal schedule $S^*_{C_{max}}$ and the C_{min}-optimal schedule $S^*_{C_{min}}$ for this instance are to be found in the following table (optimality to be seen by hand calculation).

	C_1	C_2	C_3	C_{max}	C_{min}	C_Δ
$S^*_{C_{max}} = (1, 2, 2, 3, 3, 3)$	12	15	15	15	12	3
$S^*_{C_{min}} = (1, 2, 3, 3, 2, 1)$	16	13	13	16	13	3

Table 4.3: Machine completion times and objective function values of the optimal schedules concerning $I_{C_{max},C_{min}} = (12, 8, 7, 6, 5, 4)$

Regarding the instance $I_{C_{max},C_{min}}$, note that both schedules $S^*_{C_{max}}$ and $S^*_{C_{min}}$ are also C_Δ-optimal. Hence, instances exist so that the C_Δ-optimal schedules have different makespans. We remark that $I_{C_{max},C_{min}}$ is also the smallest instance that we have found fulfilling this property.

In case of not pairwise different processing times the smallest instance that we have found is $\tilde{I}_{C_{max},C_{min}} = (9, 6, 6, 4, 4, 4)$. Regarding this instance, the C_{min}-optimal schedule is not unique, and the C_{max}-optimal schedule is the same as the one for the instance $I_{C_{max},C_{min}}$. The corresponding machine completion times (C_1, C_2, C_3) are

- $(9, 12, 12)$ of the C_{max}-optimal schedule and

- $(13, 10, 10)$ of a C_{min}-optimal schedule.

As in the sections before, with the basic example $I_{C_{max},C_{min}}$ the construction of instances with $n = m + 3$ jobs and $m \geq 4$ machines so that none of the C_{max}-optimal schedules is optimal for C_{min} is easily managed.

4.5   Joint Comparison of $C_{max}, C_{min}, C_\Delta$

This section deals with a joint comparison of the three relevant objective functions $C_{max}, C_{min}, C_\Delta$. We present the smallest instance concerning the question whether job-systems exist so that none of the C_Δ-optimal schedules is optimal for C_{max} or optimal for C_{min}. The instances presented in Section 4.2 and 4.3 do not fulfill this property as at least one of the C_Δ-optimal schedules is also optimal for C_{max} or C_{min}.

We will see that such instances exist indeed. Regarding such instances, we can also state that the set of C_{max}-optimal schedules and the set of C_{min}-optimal schedules is disjoint as the next theorem reveals. Hence, all three sets of optimal schedules are disjoint.

Theorem 4.5.1
Let S be a schedule that is both C_{max}- and C_{min}-optimal for some feasible (m, n)-instance I. Then, S is also C_Δ-optimal for I.

Proof
The proof is quite simple. Since S is C_{max}- and C_{min}-optimal we know that

$$C^*_{max} = C_{max}(S) \leq C_{max}(\tilde{S})$$
$$C^*_{min} = C_{min}(S) \geq C_{min}(\tilde{S})$$

for any schedule $\tilde{S} \neq S$. Concerning the C_Δ-value of S we can conclude:

$$C_\Delta(S) = C^*_{max} - C^*_{min} \leq C_{max}(\tilde{S}) - C_{min}(\tilde{S}) = C_\Delta(\tilde{S}).$$

This means that $C^*_\Delta = C_\Delta(S)$, i. e., S is also C_Δ-optimal.   ∎

The smallest instance with pairwise different processing times that we have found which fulfills the property that the three sets of optimal schedules are disjoint is the $(3, 7)$-instance $I_{C_{max}, C_{min}, C_\Delta} = (46, 39, 27, 26, 16, 13, 10)$. Relevant information on the C_{max}-optimal schedule $S^*_{C_{max}}$, the C_{min}-optimal schedule $S^*_{C_{min}}$, and the C_Δ-optimal schedule $S^*_{C_\Delta}$ which are unique for this instance are to be found in Table 4.4 on page 95 (optimality and uniqueness to be seen by hand calculation).

In case of not pairwise different processing times, the smallest instance that we have found is $\tilde{I}_{C_{max}, C_{min}, C_\Delta} = (45, 38, 26, 26, 16, 13, 10)$. Here, it is interesting to note that the three optimal schedules are the same as the ones for the instance $I_{C_{max}, C_{min}, C_\Delta}$. In addition, the schedule $S = (1, 2, 2, 3, 3, 3, 1)$ is C_{min}-optimal for $\tilde{I}_{C_{max}, C_{min}, C_\Delta}$, too.

	C_1	C_2	C_3	C_{max}	C_{min}	C_Δ
$S^*_{C_{max}} = (1,2,3,3,1,2,2)$	62	62	53	62	53	9
$S^*_{C_{min}} = (1,2,3,2,3,3,1)$	56	65	56	65	56	9
$S^*_{C_\Delta} = (1,2,3,3,2,1,3)$	59	55	63	63	55	8

Table 4.4: Machine completion times and objective function values of the optimal schedules concerning $I_{C_{max},C_{min},C_\Delta} = (46, 39, 27, 26, 16, 13, 10)$

Moreover, the two smallest instances $I_{C_{max},C_{min},C_\Delta} = (46, 39, 27, 26, 16, 13, 10)$ and $\tilde{I}_{C_{max},C_{min},C_\Delta} = (45, 38, 26, 26, 16, 13, 10)$ differ only in the processing times of the three longest jobs, and the difference is one each.

The following figure summarizes the smallest instances that we have found with (a lot of) computer help to contribute to the non-equivalence of the three objectives analyzed in this thesis.

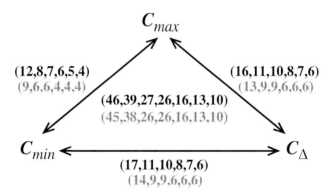

Figure 4.1: Smallest instances in case $m = 3$

4.6   Experimental Results

We finish this short chapter by presenting some results of our experimental studies on optimal scheduling in case of three machines. Again, we took the stochastic model of Section 2.5.2 (cf. page 47) and determined experimentally the probability that the C_Δ-optimal schedule is

- not C_{max}-optimal,

- not C_{min}-optimal,

- neither C_{max}- nor C_{min}-optimal.

Note that the C_Δ-optimal schedule is unique in our scenario.

Since each of the three specific IMSP's is NP-hard, we restricted our simulations to small numbers of jobs, i. e., $n \in \{5, 6, \ldots, 12\}$, and generated 10^a independent $(3, n)$-instances for each n. Again, the exponent a takes reasonable integral values (depending on n). The entries are rounded to two decimal places.

n	5	6	7	8	9	10	11	12
a	6	6	6	6	6	6	5	4
$C_{max}(S^*_{C_\Delta}) > C^*_{max}$	0.00	0.77	1.71	2.98	4.70	6.44	8.08	9.57
$C_{min}(S^*_{C_\Delta}) < C^*_{min}$	0.00	0.62	1.51	2.96	4.73	6.50	8.31	9.59
$C_{max}(S^*_{C_\Delta}) > C^*_{max}$ and $C_{min}(S^*_{C_\Delta}) < C^*_{min}$	0.00	0.00	0.00	0.02	0.06	0.16	0.27	0.45

Table 4.5: Relative number (in %) of instances so that the C_Δ-optimal schedule is not C_{max}-, C_{min}- or neither C_{max}- nor C_{min}-optimal

Obviously, all three sequences are monotonically increasing. In particular, it is interesting to find that the relative number of instances so that the C_Δ-optimal schedule is not C_{max}-optimal is almost $\frac{1}{10}$ even in case of only $n = 12$ jobs. The same is true if C_{min} is considered instead of C_{max}. Here, it would be quite interesting to know how these two sequences will continue when n is growing.

Concerning the third sequence, i. e., the one which contains information on how often the C_Δ-optimal schedule is neither C_{max}- nor C_{min}-optimal, it would also be interesting to know how this sequence continues. At least for small numbers of jobs ($n \leq 12$), the probability is greater than 0.995 that the C_Δ-optimal schedule is also optimal for at least one of the other two objectives.

Provided that the C_Δ-optimal schedule is not C_{max}-optimal (C_{min}-optimal), we also determined the relative error $\frac{C_{max}(S^*_{C_\Delta}) - C^*_{max}}{C^*_{max}}$ $\left(\frac{C^*_{min} - C_{min}(S^*_{C_\Delta})}{C^*_{min}} \right)$. The

n	5	6	7	8	9	10	11	12
a	6	6	6	6	6	6	5	4
$\frac{C_{max}(S^*_{C_\Delta})-C^*_{max}}{C^*_{max}}$	0	0.015	0.010	0.006	0.004	0.002	0.001	0.001
$\frac{C^*_{min}-C_{min}(S^*_{C_\Delta})}{C^*_{min}}$	0	0.019	0.011	0.006	0.004	0.002	0.001	0.001

Table 4.6: Relative errors $\frac{C_{max}(S^*_{C_\Delta})-C^*_{max}}{C^*_{max}}$ and $\frac{C^*_{min}-C_{min}(S^*_{C_\Delta})}{C^*_{min}}$ provided that the C_Δ-optimal schedule is not C_{max}-optimal (C_{min}-optimal)

results rounded to three decimal places are contained in Table 4.6. Disregarding the entries in the $n = 5$-column, the relative errors are monotonic decreasing and they are almost zero even for such small numbers of jobs. We conjecture that the relative errors will further decrease when n is growing.

To summarize our few experimental results of this section, we conjecture that the C_Δ-optimal is not C_{max}-optimal (C_{min}-optimal) with high probability for large n. However, the relative errors of the corresponding objective function values are almost zero in such cases.

Chapter 5

Conclusions and Open Problems

This chapter summarizes the main results and observations of this thesis. It also discusses ideas and open questions for future work on identical machine scheduling problems.

5.1 Conclusions

The aim of this thesis was to analyze various aspects of scheduling independent jobs on identical machines. We considered three different objectives: minimizing the maximum completion time C_{max} (the makespan), maximizing the minimum completion time C_{min}, and minimizing the difference $C_\Delta = C_{max} - C_{min}$.

Besides investigations on heuristic algorithms we contributed to optimal scheduling although all three specific scheduling problems are NP-hard.

Our studies on heuristic algorithms were mainly concentrated on List Scheduling (LS). With this procedure, jobs are arranged in some arbitrary order (specified by a priority rule). Whenever a machine becomes free for processing, LS removes the first job on the list and assigns it to that machine.

One goal was to compare two of the most popular priority rules of the LS-algorithm: the SPT- (shortest processing time) and the LPT-rule (longest processing time). So, we analyzed the structure of SPT-LS-schedules. We found out that with the SPT-rule, every m-th job of the list is assigned to the same machine (and no other jobs are assigned to that machine). With this result and simple bounds on the maximum as well as the minimum completion time of LS-schedules, we were able to prove for all three objective

functions that the SPT-rule is dominated by the LPT-rule. Regarding any
of the three objective functions, this means that no instance exists so that the
SPT-rule yields a better value than the LPT-rule, and at least one instance
exists so that the SPT-rule yields a worse value than the LPT-rule.
Investigations concerning dominance were one of the main research interests
of this thesis. Therefore, a deeper look at the SPT-rule revealed that the
SPT-rule is not the worst priority rule of the LS-algorithm. For each of the
three objective functions, we presented a class of priority lists which is dominated by the SPT-rule.
Another important result in connection with dominance concerns the RLPT-
heuristic (restricted largest processing time) which is based on the LPT-rule.
Here, we were able to prove that the LPT-rule dominates the RLPT-heuristic
in terms of C_{min}-maximization and C_Δ-minimization. The dominance of the
LPT-rule in terms of C_{max}-minimization has already been known. Furthermore, we showed that the RLPT-heuristic is dominating the SPT-rule for all
three objectives.
We also studied the worst-case performance of the RLPT-heuristic relative to
optimal scheduling as well as to LPT-scheduling precisely. We showed that
the makespan of the RLPT-schedule is never more than $(2 - \frac{1}{m})$ times the
optimal makespan and that the minimum completion time is at least $\frac{1}{m}$ of
the optimal minimum completion time. We proved the performance bounds
to be asymptotically tight and showed that no instances exist so that the
bounds are reached exactly. It was interesting to find that the same performance bounds apply when RLPT-scheduling is compared to LPT-scheduling
instead of optimal scheduling.

Regarding optimal scheduling, we were also interested in various aspects.
Main attention was directed on potentially optimal schedules in case of two
machines, i. e., schedules for which feasible processing times exist so that
these schedules are optimal. We introduced a transformation of schedules to
one-dimensional paths whereby we gave a total characterization of the set of
potentially optimal schedules.
We also studied whether schedules are optimal on convex sets in the polytope
$1 \geq t_1 \geq \ldots \geq t_n \geq 0$ of feasible processing times. Even in case of two
machines we found instances which do not fulfill the convexity-property. In
contrast to this, List Scheduling (assuming any fixed priority rule) fulfills the
convexity-property.

Concerning optimal scheduling in case of at least three machines, we contributed to the non-equivalence of the three objective functions under investigation and presented smallest instances concerning this matter. We com-

pared each pair of objective functions, and we also did a joint comparison of all three objective functions.

This thesis also contains some experimental results. In our stochastic model we assumed the processing times to be independent samples, uniformly distributed in the unit interval $[0, 1]$.

In case of two machines we found out that the LPT-schedule is shorter than the RLPT-schedule with probability $\frac{1}{4}$ (n even, $n \geq 4$) or $\frac{1}{8}$ (n odd, $n \geq 5$). Concerning optimal scheduling in case of three machines and small numbers of jobs ($n \leq 12$), we determined experimentally the probability that the C_Δ-optimal schedule is not C_{max}-optimal (C_{min}-optimal). The two sequences of the corresponding probabilities are strictly monotonic increasing in n. In case $n = 12$ the probability is approximately $\frac{1}{10}$ each. However, the average relative errors $\frac{C_{max}(S^*_{C_\Delta}) - C^*_{max}}{C^*_{max}}$ and $\frac{C^*_{min} - C_{min}(S^*_{C_\Delta})}{C^*_{min}}$ are approximately zero in such cases - even for small values of n.

5.2   Open Problems and Future Work

We finish this thesis by stating a collection of ideas, approaches and open questions for future work. The order in which the ideas are listed below is related to the structure of this thesis.

1. In Section 1.3, we presented the objective functions C_{max}, C_{min} and $C_\Delta = C_{max} - C_{min}$. In case of three machines, one can generalize them for instance to the objective of minimizing $C_{(\lambda_3, \lambda_2, \lambda_1)} = \lambda_3 C_{[3]} + \lambda_2 C_{[2]} + \lambda_1 C_{[1]}$ ($\lambda_3 \geq \lambda_2 \geq \lambda_1$)[1]. Due to the equality $\sum_{i=1}^{3} C_{[i]} = \sum_{j=1}^{n} t_j$, minimizing $C_{(\lambda_3, \lambda_2, \lambda_1)}$ is equivalent to minimizing $\tilde{C}_{(\lambda_3, \lambda_2, \lambda_1)} = (\lambda_3 - \lambda_2)C_{[3]} - (\lambda_2 - \lambda_1)C_{[1]}$.

   W. l. o. g. one can choose $\lambda_2 = 0$, and one can claim equality $\lambda_3 + |\lambda_1| = 1$. This leads to $C_\lambda = \lambda C_{[3]} - (1 - \lambda)C_{[1]}$ ($\lambda \subset [0, 1]$) which is to be minimized. Hence, $\lambda = 1$ is equivalent to the minimization of C_{max}, $\lambda = 0$ is equivalent to the maximization of C_{min}, and $\lambda = \frac{1}{2}$ is equivalent to the minimization of C_Δ.

   On the one hand, it seems to be interesting to analyze which priority rules work well for fixed $\lambda \in [0, 1]$. On the other hand, which is the best λ for a fixed priority rule?

---

[1]Note that $C_{[i]}$ denotes the completion time of the i-th shortest running machine (cf. page 23).

2. In Section 2.5.1 we were interested in the question whether priority lists exist which are dominated by the SPT-LS-rule. In this connection, it would be interesting to know if priority lists exist which dominate the LPT-rule.

3. In Chapter 2 we stated four conjectures (Conjecture 2.5.20, 2.6.8, 2.6.9 and 2.6.13 on the pages 50, 64, 65 and 67, respectively) for which we do not have theoretical proofs so far. Proofing (or falsifying) at least one of these conjectures seems to be a challenging task.

4. Analogously to the case of two machines, does there exist an elegant way to characterize the set of potentially optimal C_{max}-, C_{min}- and C_Δ-optimal schedules for any fixed number $m \geq 3$ of machines?

5. On page 83 we conjectured that each potentially optimal schedule is also potentially unique optimal in case of two machines. How can the construction schemes of the processing times presented in the proof of Theorem 3.2.5 be modified so that the corresponding schedules are uniquely optimal?

6. In his doctoral thesis, Doernfelder [Doe09] investigated the maximum number of threshold parameters of penalty methods in discrete optimization. In this connection, the following question concerns the non-convexity-property of C_{max}-optimal scheduling presented in Section 3.3.
   So, assume schedule S to be C_{max}-optimal for two different instances I_1 and I_2 and consider all schedules which are optimal for any convex combined instance $I_\lambda = \lambda I_1 + (1 - \lambda)I_2$ $(0 < \lambda < 1)$. Then, one can ask for the maximum number of different λ-values $(\lambda \in (0,1))$ at which the C_{max}-optimal solution switches from one schedule to another schedule. The analysis can be done depending on m and n as well as for the other two objective functions C_{min} and C_Δ.

7. In Chapter 4, we presented smallest instances assuming a certain lexicographic order (cf. page 90) to contribute to the non-equivalence of the three objective functions C_{max}, C_{min}, C_Δ in case of at least three machines. Considering other lexicographic orders, how do smallest instances look like then?

8. Concerning (lexicographically) optimal scheduling, which are the worst-case bounds of the following ratios for any fixed $m \geq 3$:
$$\frac{C_{max}(S^*_{C_{min}})}{C^*_{max}}, \quad \frac{C_{max}(S^*_{C_\Delta})}{C^*_{max}}, \quad \frac{C_{min}(S^*_{C_{max}})}{C^*_{min}}, \quad \frac{C_{min}(S^*_{C_\Delta})}{C^*_{min}} ?$$

Thereby, $S^*_{C_{max}}$ represents a lexicographically (C_{max}, C_{min})-optimal schedule, $S^*_{C_{min}}$ represents a lexicographically (C_{min}, C_{max})-optimal schedule, and depending on which ratio is considered, $S^*_{C_\Delta}$ represents either a lexicographically (C_Δ, C_{max})- or (C_Δ, C_{min})-optimal schedule.

9. In connection with the previous item, which of the two lexicographically optimal schedules $S^*_{C_{min}}$ and S^*_Δ performs on average better concerning the objective function C_{max}? Analogously, this question can be studied for the other two objective functions.

10. According to our experimental results on the probability that the C_Δ-optimal schedule is not C_{max}-optimal (C_{min}-optimal) in case $m = 3$ presented in Section 4.6, are the two corresponding sequences of probabilities converging to one in n?

Appendix A

Heuristics - Comparing LPT and RLPT in Case of Two Machines

The appendix is primarily meant to prove the Lemmas 2.6.11 and 2.6.12 of Section 2.6.3.3. Therefore, we initially give an alternative proof in case of two machines of the result by Coffman and Sethi [CS76] on the length of the LPT- and the RLPT-schedule. Then, in Appendix A.2 we use the results of our proof to show the correctness of the two lemmas mentioned before. Eventually, Appendix A.3 contains some of our experimental results concerning Conjecture 2.6.13 of page 67.

A.1 An Alternative Proof of the Dominance of LPT over RLPT

Theorem A.1.1
Consider an arbitrary $(2, n)$-instance of the IMSP. Then, the RLPT-schedule is at least as long as the LPT-schedule, i. e., $C_{max}^{RLPT} \geq C_{max}^{LPT}$.

Proof of the Theorem:
The idea of the proof is to consider the jobs pairwise like in RLPT. After the assignment of $2j$ jobs both heuristics have constructed a partial schedule either with the same makespan or with different makespans. In case of different makespans we will see that the makespan of the LPT-schedule is better than the makespan of the RLPT-schedule at that point.
For both cases described before we will then show that the assignment of the

remaining $(n - 2j)$ jobs cannot lead to a situation in which the makespan of the LPT-schedule is worse than the makespan of the RLPT-schedule.

As in Section 2.6.3.3, j^* denotes the job after whose assignment $C_{max}^{LPT}(j^*) < C_{max}^{RLPT}(j^*)$ holds for the first time (if such a job exists).

Most of the proof below is rather straightforward. The crucial points are the condition $C_\Delta^{LPT}(2j) > t_{2j+1}$ and the statement (ii) in Lemma A.1.4, where the *max*-term of heuristic RLPT is compared with the *min*-term of LPT.

Lemma A.1.2
Let $C_{max}^{LPT}(2j) = C_{max}^{RLPT}(2j)$ for some $j \in \{1, \ldots, \frac{n}{2} - 1\}$, then $C_{max}^{LPT}(2j+2) \leq C_{max}^{RLPT}(2j+2)$.

Proof of Lemma A.1.2:
According to RLPT, the machine completion times after the assignment of $(2j + 2)$ jobs are $C_{max}^{RLPT}(2j) + t_{2j+2}$ and $C_{min}^{RLPT}(2j) + t_{2j+1}$.
According to LPT, the assignment of job $(2j + 2)$ depends on which machine has minimum completion time after the assignment of $(2j + 1)$ jobs. Two cases are possible:

Case 1: $C_\Delta^{LPT}(2j) \leq t_{2j+1}$

This leads to the same situation as in the RLPT-schedule, namely one machine has completion time $C_{max}^{LPT}(2j) + t_{2j+2}$ and the other machine has completion time $C_{min}^{LPT}(2j) + t_{2j+1}$ after the assignment of $(2j + 2)$ jobs according to LPT. Hence $C_{max}^{LPT}(2j + 2) = C_{max}^{RLPT}(2j + 2)$.

Case 2: $C_\Delta^{LPT}(2j) > t_{2j+1}$

This leads to a situation in which one machine has completion time $C_{max}^{LPT}(2j)$ and the other machine has completion time $C_{min}^{LPT}(2j) + t_{2j+1} + t_{2j+2}$ after the assignment of $(2j + 2)$ jobs according to LPT. If $t_{2j+2} = 0$ we have again $C_{max}^{LPT}(2j + 2) = C_{max}^{RLPT}(2j + 2)$. If $t_{2j+2} > 0$ we have $C_{max}^{LPT}(2j + 2) < C_{max}^{RLPT}(2j + 2)$.

This completes the proof of Lemma A.1.2. □

Using Lemma A.1.2 and the fact that equality $C_{max}^{LPT}(j) = C_{max}^{RLPT}(j)$ holds for the starting parameters $j = 1, 2$ it follows immediately that after the assignment of an odd number of jobs it cannot occur that the makespan of the LPT-schedule differs from the makespan of the RLPT-schedule for the first time. Hence j^* has to be even and $t_{j^*} > 0$.

Lemma A.1.3

Let $C_\Delta^{RLPT}(2j) > t_{2j+1}$ for some $j \in \{1, \ldots, \frac{n}{2} - 1\}$. Then

(i) $C_{max}^{RLPT}(2j + 2) = C_{max}^{RLPT}(2j) + t_{2j+2}$

(ii) $C_\Delta^{RLPT}(2j + 2) > t_{2j+2}$.

Proof of Lemma A.1.3:

Assume $C_\Delta^{RLPT}(2j) > t_{2j+1}$ for some $j \in \{1, \ldots, \frac{n}{2} - 1\}$. This means that $C_{min}^{RLPT}(2j) + t_{2j+1} < C_{max}^{RLPT}(2j)$. So, we get $C_{max}^{RLPT}(2j + 1) = C_{max}^{RLPT}(2j)$. RLPT assigns job $(2j + 2)$ to the machine with completion time $C_{max}^{RLPT}(2j)$. This results in $C_{max}^{RLPT}(2j + 2) = C_{max}^{RLPT}(2j) + t_{2j+2}$ and

$$C_\Delta^{RLPT}(2j + 2) = C_{max}^{RLPT}(2j + 2) - C_{min}^{RLPT}(2j + 2) =$$
$$= C_{max}^{RLPT}(2j) + t_{2j+2} - \left(C_{min}^{RLPT}(2j) + t_{2j+1}\right) >$$
$$> C_{max}^{RLPT}(2j) + t_{2j+2} - C_{max}^{RLPT}(2j) = t_{2j+2}.$$

This completes the proof of Lemma A.1.3.   □

By Lemma A.1.3 we know that $C_{max}^{RLPT}(2j) = C_{max}^{RLPT}(2j - 2) + t_{2j}$ for all $j = \frac{t^*}{2}, \ldots, \frac{n}{2}$, if a given $(2, n)$-instance contains a job j^*.

Lemma A.1.4

Let $C_{max}^{LPT}(2j) \leq C_{max}^{RLPT}(2j)$ for some $j \in \{1, \ldots, \frac{n}{2} - 1\}$ and $C_\Delta^{LPT}(2j) > t_{2j+1}$. Then

(i) $C_{max}^{RLPT}(2j + 2) = C_{max}^{RLPT}(2j) + t_{2j+2}$ and $C_\Delta^{RLPT}(2j + 2) > t_{2j+2}$

(ii) $C_{max}^{RLPT}(2j + 2) - C_{min}^{LPT}(2j + 2) \geq t_{2j+2}$.

Proof of Lemma A.1.4:

The condition $C_{max}^{LPT}(2j) \leq C_{max}^{RLPT}(2j)$ for some $j \in \{1, \ldots, \frac{n}{2} - 1\}$ is equivalent to $C_{min}^{LPT}(2j) \geq C_{min}^{RLPT}(2j)$. This condition and condition $C_\Delta^{LPT}(2j) > t_{2j+1}$ ensure $C_\Delta^{RLPT}(2j) > t_{2j+1}$, too. Then, statement (i) follows directly from Lemma A.1.3.

$C_\Delta^{LPT}(2j) > t_{2j+1}$ means that LPT assigns the jobs $(2j + 1)$ and $(2j + 2)$ both to the machine with minimum completion time after the assignment of $2j$ jobs. To verify statement (ii), consider the following two cases.

Case 1: The **max**-machine after $2j$ jobs remains **max**-machine after $(2j+2)$ jobs, i. e., $C_{max}^{LPT}(2j+2) = C_{max}^{LPT}(2j) > C_{min}^{LPT}(2j+2)$.

Using statement (i) of Lemma A.1.3 yields

$$C_{max}^{RLPT}(2j+2) - C_{min}^{LPT}(2j+2) > C_{max}^{RLPT}(2j+2) - C_{max}^{LPT}(2j+2) =$$
$$= C_{max}^{RLPT}(2j) + t_{2j+2} - C_{max}^{LPT}(2j) \geq t_{2j+2}.$$

Case 2: The **min**-machine after $2j$ jobs becomes **max**-machine after $(2j+2)$ jobs, i. e., $C_{max}^{LPT}(2j+2) = C_{min}^{LPT}(2j) + t_{2j+1} + t_{2j+2}$.

Using statement (i) of Lemma A.1.3 once again yields

$$C_{max}^{RLPT}(2j+2) - C_{min}^{LPT}(2j+2) = C_{max}^{RLPT}(2j) + t_{2j+2} - C_{max}^{LPT}(2j) \geq t_{2j+2}.$$

This completes the proof of Lemma A.1.4. □

Lemma A.1.4 ensures $C_{max}^{RLPT}(j^*) - C_{min}^{LPT}(j^*) \geq t_{j^*}$, if a given $(2, n)$-instance contains a job j^*.

The next lemma is something like the step of induction, transferring statements for $2j$ to statements for $(2j + 2)$.

Lemma A.1.5
If $C_{max}^{LPT}(2j) \leq C_{max}^{RLPT}(2j)$ and $C_{max}^{RLPT}(2j) - C_{min}^{LPT}(2j) \geq t_{2j}$ for some $j \in \{1, \ldots, \frac{n}{2} - 1\}$, then $C_{max}^{LPT}(2j + 2) \leq C_{max}^{RLPT}(2j + 2)$ and $C_{max}^{RLPT}(2j + 2) - C_{min}^{LPT}(2j + 2) \geq t_{2j+2}$.

Proof of Lemma A.1.5:
Assume $C_{max}^{LPT}(2j) \leq C_{max}^{RLPT}(2j)$ and $C_{max}^{RLPT}(2j) - C_{min}^{LPT}(2j) \geq t_{2j}$ for some $j \in \{1, \ldots, \frac{n}{2} - 1\}$. After the assignment of job $(2j + 1)$ the machine completion times of the partial RLPT-schedule are $C_{max}^{RLPT}(2j)$ and $C_{min}^{RLPT}(2j) + t_{2j+1}$, whereas the machine completion times of the partial LPT-schedule are $C_{max}^{LPT}(2j)$ and $C_{min}^{LPT}(2j) + t_{2j+1}$. Since $t_{2j+1} \leq t_{2j}$, it follows that $C_{min}^{LPT}(2j) + t_{2j+1} \leq C_{max}^{RLPT}(2j) = C_{max}^{RLPT}(2j + 1)$. Moreover, we have $C_{max}^{LPT}(2j + 1) \leq C_{max}^{RLPT}(2j + 1)$ and

$$C_{max}^{RLPT}(2j + 2) = C_{max}^{RLPT}(2j) + t_{2j+2} =$$
$$= C_{max}^{RLPT}(2j + 1) + t_{2j+2} \geq C_{max}^{LPT}(2j + 1) + t_{2j+2} \geq C_{max}^{LPT}(2j + 2).$$

To verify that the inequality $C_{max}^{RLPT}(2j + 2) - C_{min}^{LPT}(2j + 2) \geq t_{2j+2}$ is true, consider the following two cases.

Case 1: The **min**-machine after $(2j + 1)$ jobs remains **min**-machine after $(2j + 2)$ jobs, i. e., $C_{min}^{LPT}(2j + 2) = C_{min}^{LPT}(2j + 1) + t_{2j+2} < C_{max}^{LPT}(2j + 1)$. Then, we can conclude that

$$C_{max}^{RLPT}(2j+2) - C_{min}^{LPT}(2j+2) > C_{max}^{RLPT}(2j) + t_{2j+2} - C_{max}^{LPT}(2j+1) \geq t_{2j+2}.$$

Case 2: The **max**-machine after $(2j + 1)$ jobs becomes **min**-machine after $(2j + 2)$ jobs, i. e., $C_{min}^{LPT}(2j + 2) = C_{max}^{LPT}(2j + 1)$. Then, using $C_{max}^{RLPT}(2j + 1) = C_{max}^{RLPT}(2j)$ yields

$$C_{max}^{RLPT}(2j + 2) - C_{min}^{LPT}(2j + 2) = C_{max}^{RLPT}(2j + 2) - C_{max}^{LPT}(2j + 1) =$$
$$= C_{max}^{RLPT}(2j + 1) + t_{2j+2} - C_{max}^{LPT}(2j + 1) \geq t_{2j+2}.$$

This completes the proof of Lemma A.1.5. $\square$

For $(2, n)$-instances in which a job j^* exists, Lemma A.1.5 ensures the desired property that $C_{max}^{LPT}(j) \leq C_{max}^{RLPT}(j)$ for all $j = j^* + 1, \ldots, n$.

If no job j^* exists, equality $C_{max}^{LPT}(j) = C_{max}^{RLPT}(j)$ is true for all $j = 1, \ldots, n$ of course.

This completes the proof of Theorem A.1.1. $\blacksquare$

A.2 Proofs of the Lemmas 2.6.11 and 2.6.12

In a next step we want to prove the Lemmas 2.6.11 and 2.6.12. Therefore, we use the result of Lemma A.1.4 which ensures $C_{max}^{RLPT}(j^*) - C_{min}^{LPT}(j^*) \geq t_{j^*}$ if a given $(2, n)$-instance contains a job j^*.

Instead of proofing the two lemmas directly, we formulate two related lemmas and prove them. Then, the proofs of the Lemmas 2.6.11 and 2.6.12 are easily managed by transferring the results of the following two lemmas analogously to Lemma A.1.4 and Lemma A.1.5.

Lemma A.2.1 deals with the case that $C_{max}^{RLPT}(j^*) - C_{min}^{LPT}(j^*) > t_{j^*}$ and is related to Lemma 2.6.11, whereas Lemma A.2.2 considers the other case that $C_{max}^{RLPT}(j^*) - C_{min}^{LPT}(j^*) = t_{j^*}$. Lemma A.2.2 is related to Lemma 2.6.12.

Lemma A.2.1

Consider an arbitrary $(2, n)$-instance of the IMSP that contains a job j^* $(j^* < n)$. If $C_{max}^{RLPT}(j^*) - C_{min}^{LPT}(j^*) > t_{j^*}$, then

(i) $C_{max}^{LPT}(j) < C_{max}^{RLPT}(j)$ for $j = j^* + 1, j^* + 2$ and

(ii) $C_{max}^{RLPT}(j^* + 2) - C_{min}^{LPT}(j^* + 2) > t_{j^*+2}$.

Proof

As we consider a $(2, n)$-instance which contains a job j^*, we know by the definition of j^* that $C_{max}^{LPT}(j^*) < C_{max}^{RLPT}(j^*)$. Furthermore, we consider the case that $C_{max}^{RLPT}(j^*) - C_{min}^{LPT}(j^*) > t_{j^*}$.

In order to prove the two statements of Lemma A.2.1, we take a look at the makespans of the (partial) LPT- and RLPT-schedule after the assignment of $(j^* + 1)$ and $(j^* + 2)$ jobs.

Concerning the RLPT-schedule we know by Lemma A.1.3 that $C_{max}^{RLPT}(j^* + 1) = C_{max}^{RLPT}(j^*)$ and $C_{max}^{RLPT}(j^* + 2) = C_{max}^{RLPT}(j^*) + t_{j^*+2}$.

Concerning the LPT-schedule we know that

$$C_{max}^{LPT}(j^* + 1) \in \{C_{max}^{LPT}(j^*), C_{min}^{LPT}(j^*) + t_{j^*+1}\},$$

which directly leads to $C_{max}^{LPT}(j^* + 1) < C_{max}^{RLPT}(j^* + 1)$, as $C_{max}^{LPT}(j^*) < C_{max}^{RLPT}(j^*)$ and $C_{min}^{LPT}(j^*) + t_{j^*+1} \leq C_{min}^{LPT}(j^*) + t_{j^*} < C_{max}^{RLPT}(j^*)$.

We can bound the **max**-term of the LPT-schedule after $(j^* + 2)$ jobs by

$$C_{max}^{LPT}(j^*+2) \leq C_{max}^{LPT}(j^*+1)+t_{j^*+2} < C_{max}^{RLPT}(j^*+1)+t_{j^*+2} = C_{max}^{RLPT}(j^*+2)$$

and the **min**-term by

$$C_{min}^{LPT}(j^* + 2) \leq C_{max}^{LPT}(j^* + 1) < C_{max}^{RLPT}(j^* + 1).$$

Thus, $C_{max}^{RLPT}(j^*+2) - C_{min}^{LPT}(j^*+2) > C_{max}^{RLPT}(j^*+2) - C_{max}^{RLPT}(j^*+1) = t_{j^*+2}$. This completes the proof of the two statements of Lemma A.2.1. ∎

Lemma A.2.2

Consider an arbitrary $(2, n)$-instance of the IMSP that contains a job j^* $(j^* < n)$. Furthermore, let $C_{max}^{RLPT}(j^*) - C_{min}^{LPT}(j^*) = t_{j^*}$.

(i) If $t_{j^*} > t_{j^*+1}$, then $C_{max}^{LPT}(j) < C_{max}^{RLPT}(j)$ for $j = j^* + 1, j^* + 2$ and $C_{max}^{RLPT}(j^* + 2) - C_{min}^{LPT}(j^* + 2) = t_{j^*+2}$.

(ii) If $t_{j^*} = t_{j^*+1}$, then $C_{max}^{LPT}(j^* + 1) = C_{max}^{RLPT}(j^* + 1)$.

(iii) If $t_{j^*} = t_{j^*+1} = t_{j^*+2}$, then $C_{max}^{LPT}(j^* + 2) < C_{max}^{RLPT}(j^* + 2)$ and $C_{max}^{RLPT}(j^* + 2) - C_{min}^{LPT}(j^* + 2) = t_{j^*+2}$.

(iv) If $t_{j^*} = t_{j^*+1} > t_{j^*+2} > 0$, then $C_{max}^{LPT}(j^*+2) < C_{max}^{RLPT}(j^*+2)$ and

$$C_{max}^{RLPT}(j^*+2) - C_{min}^{LPT}(j^*+2) \begin{cases} = t_{j^*+2} & \text{if } C_{min}^{LPT}(j^*+2) = C_{max}^{LPT}(j^*+1), \\ > t_{j^*+2} & \text{if } C_{min}^{LPT}(j^*+2) < C_{max}^{LPT}(j^*+1). \end{cases}$$

Proof

As we consider a $(2, n)$-instance which contains a job j^*, we know by the definition of j^* that $C_{max}^{LPT}(j^*) < C_{max}^{RLPT}(j^*)$. Furthermore, we consider the case that $C_{max}^{RLPT}(j^*) - C_{min}^{LPT}(j^*) = t_{j^*}$.

To prove statement (i), we take a look at the makespan of the (partial) LPT- and RLPT-schedule after the assignment of (j^*+1) and (j^*+2) jobs. The proof of this statement works analogously to the proof of Lemma A.2.1.

Concerning the RLPT-schedule, we get $C_{max}^{RLPT}(j^*+1) = C_{max}^{RLPT}(j^*)$ and $C_{max}^{RLPT}(j^*+2) = C_{max}^{RLPT}(j^*) + t_{j^*+2}$ by Lemma A.1.3.

Concerning the LPT-schedule we know that

$$C_{max}^{LPT}(j^*+1) \in \{C_{max}^{LPT}(j^*), C_{min}^{LPT}(j^*) + t_{j^*+1}\},$$

which directly leads to $C_{max}^{LPT}(j^*+1) < C_{max}^{RLPT}(j^*+1) = C_{max}^{RLPT}(j^*)$ as $C_{max}^{LPT}(j^*) < C_{max}^{RLPT}(j^*)$ and $t_{j^*} > t_{j^*+1}$.

Analogously to the proof of Lemma A.2.1, the **max**-term of the LPT-schedule after (j^*+2) jobs can be bounded by

$$C_{max}^{LPT}(j^*+2) \le C_{max}^{LPT}(j^*+1) + t_{j^*+2} < C_{max}^{RLPT}(j^*+1) + t_{j^*+2} = C_{max}^{RLPT}(j^*+2)$$

and the **min**-term by

$$C_{min}^{LPT}(j^*+2) \le C_{max}^{LPT}(j^*+1) < C_{max}^{RLPT}(j^*+1).$$

Thus, $C_{max}^{RLPT}(j^*+2) - C_{min}^{LPT}(j^*+2) > C_{max}^{RLPT}(j^*+2) - C_{max}^{RLPT}(j^*+1) = t_{j^*+2}$. This completes the proof of statement (i).

The proof of statement (ii) is shorter. As

$$C_{max}^{RLPT}(j^*) = C_{min}^{LPT}(j^*) + t_{j^*} = C_{min}^{LPT}(j^*) + t_{j^*+1} > C_{max}^{LPT}(j^*),$$

we can simply conclude:

$$C_{max}^{LPT}(j^*+1) = C_{min}^{LPT}(j^*) + t_{j^*+1} = C_{min}^{LPT}(j^*) + t_{j^*} =$$
$$= C_{max}^{RLPT}(j^*) \underset{\underset{C_{min}^{RLPT}(j^*) < C_{min}^{LPT}(j^*)}{\uparrow}}{=} C_{max}^{RLPT}(j^*+1).$$

This completes the proof of statement (ii).

Next, we prove statement (iii). Therefore, we use the result of statement (ii) by which we know the current completion times of the LPT- and the RLPT-schedule after the assignment of $(j^* + 1)$ jobs. With this result, we derive the (current) machine completion times of the LPT-schedule after the assignment of $(j^* + 2)$ jobs.

Regarding the makespan of the RLPT-schedule after $(j^* + 2)$ jobs assigned we know by Lemma A.1.3 that $C_{max}^{RLPT}(j^* + 2) = C_{max}^{RLPT}(j^*) + t_{j^*+2}$.

Concerning the LPT-schedule, the (current) machine completion times after the assignment of $(j^* + 2)$ jobs are $C_{max}^{LPT}(j^*) + t_{j^*+2}$ and $C_{min}^{LPT}(j^*) + t_{j^*+1} = C_{max}^{RLPT}(j^*)$. As $C_{max}^{LPT}(j^*) \geq C_{min}^{LPT}(j^*)$ and $t_{j^*+1} = t_{j^*+2}$ we can conclude that

$$C_{max}^{LPT}(j^* + 2) = C_{max}^{LPT}(j^*) + t_{j^*+2} < C_{max}^{RLPT}(j^*) + t_{j^*+2} = C_{max}^{RLPT}(j^* + 2).$$

Furthermore, we get

$$C_{max}^{RLPT}(j^* + 2) - C_{min}^{LPT}(j^* + 2) = C_{max}^{RLPT}(j^*) + t_{j^*+2} - C_{min}^{LPT}(j^*) - t_{j^*+1} =$$
$$= \underset{\underset{t_{j^*+1} = t_{j^*+2}}{\uparrow}}{C_{max}^{RLPT}(j^*) - C_{min}^{LPT}(j^*) = t_{j^*} = t_{j^*+2}.}$$

This completes the proof of statement (iii).

In a last step we prove statement (iv). Again, we use the result of statement (ii) to derive the (current) machine completion times of the LPT-schedule after the assignment of $(j^* + 2)$ jobs.

Considering the RLPT-schedule, we already know that $C_{max}^{RLPT}(j^* + 2) = C_{max}^{RLPT}(j^*) + t_{j^*+2}$.

Regarding the makespan of the LPT-schedule after the assignment of (j^*+2) jobs, we can conclude that

$$C_{max}^{LPT}(j^* + 2) \in \{C_{max}^{LPT}(j^*) + t_{j^*+2}, C_{min}^{LPT}(j^*) + t_{j^*+1}\}.$$

As

$$C_{max}^{LPT}(j^*) + t_{j^*+2} < C_{max}^{RLPT}(j^*) + t_{j^*+2} = C_{max}^{RLPT}(j^* + 2)$$

and

$$C_{min}^{LPT}(j^*) + t_{j^*+1} = C_{max}^{RLPT}(j^*) < C_{max}^{RLPT}(j^*) + t_{j^*+2} = C_{max}^{RLPT}(j^* + 2)$$

we can conclude that $C_{max}^{LPT}(j^* + 2) < C_{max}^{RLPT}(j^* + 2)$.

Regarding the difference $C_{max}^{RLPT}(j^* + 2) - C_{min}^{LPT}(j^* + 2)$, we distinguish the following two cases. Note that no other cases can occur.

Case 1: $C_{min}^{LPT}(j^* + 2) = C_{max}^{LPT}(j^* + 1)$.

In this case, the difference is

$$C_{max}^{RLPT}(j^* + 2) - C_{min}^{LPT}(j^* + 2) = C_{max}^{RLPT}(j^* + 2) - C_{max}^{LPT}(j^* + 1) =$$
$$= C_{max}^{RLPT}(j^*) + t_{j^*+2} - (C_{min}^{LPT}(j^*) + t_{j^*+1}) =$$
$$= C_{max}^{RLPT}(j^*) + t_{j^*+2} - C_{max}^{RLPT}(j^*) = t_{j^*+2}.$$

Case 2: $C_{min}^{LPT}(j^* + 2) < C_{max}^{LPT}(j^* + 1)$.

Here, we can conclude that

$$C_{max}^{RLPT}(j^* + 2) - C_{min}^{LPT}(j^* + 2) = C_{max}^{RLPT}(j^*) + t_{j^*+2} - C_{min}^{LPT}(j^* + 2) >$$
$$> C_{max}^{RLPT}(j^*) + t_{j^*+2} - C_{max}^{LPT}(j^* + 1) =$$
$$= C_{max}^{RLPT}(j^*) + t_{j^*+2} - C_{max}^{RLPT}(j^*) = t_{j^*+2}.$$

This completes the proof of statement (iv). Hence, the proof of Lemma A.2.2 is completed. ∎

A.3 Experimental Results Concerning Conjecture 2.6.13

In this section we want to present a few experimental results in connection with Conjecture 2.6.13 of page 67. The results are meant to support this conjecture.

The following tables contain information on the relative number (in %) of instances so that the LPT- and the RLPT-heuristic generate the same schedule after the assignment of $(n - k)$ jobs. Each entry in the third line of these tables is based on 10^7 independent $(2, n)$-instances. The fourth line in Table A.1 and Table A.4 serves as a comparison. All entries are rounded to three decimal places.

$n = 16$				
k	0	2	4	6
$S^{LPT}(1:n-k) = S^{RLPT}(1:n-k)$	75.000	93.745	98.432	99.608
$(1 - \frac{1}{2^{k+2}}) \times 100$	75	93.75	98.436	99.609

$n = 16$			
k	8	10	12
$S^{LPT}(1:n-k) = S^{RLPT}(1:n-k)$	99.903	99.975	99.994
$(1 - \frac{1}{2^{k+2}}) \times 100$	99.902	99.976	99.994

Table A.1: Relative number (in %) of instances so that
$S^{LPT}(1:n-k) = S^{RLPT}(1:n-k)$ for $n = 16$ and $k \in \{0, 2, 4, 6, 8, 10, 12\}$

$n = 50$				
k	0	2	4	6
$S^{LPT}(1:n-k) = S^{RLPT}(1:n-k)$	75.005	93.750	98.436	99.611

$n = 50$			
k	8	10	12
$S^{LPT}(1:n-k) = S^{RLPT}(1:n-k)$	99.902	99.975	99.995

Table A.2: Relative number (in %) of instances so that
$S^{LPT}(1:n-k) = S^{RLPT}(1:n-k)$ for $n = 50$ and $k \in \{0, 2, 4, 6, 8, 10, 12\}$

$n = 100$				
k	0	2	4	6
$S^{LPT}(1:n-k) = S^{RLPT}(1:n-k)$	75.022	93.752	98.440	99.605

$n = 100$			
k	8	10	12
$S^{LPT}(1:n-k) = S^{RLPT}(1:n-k)$	99.903	99.976	99.994

Table A.3: Relative number (in %) of instances so that
$S^{LPT}(1:n-k) = S^{RLPT}(1:n-k)$ for $n = 100$ and $k \in \{0, 2, 4, 6, 8, 10, 12\}$

$n = 17$				
k	1	3	5	7
$S^{LPT}(1:n-k) = S^{RLPT}(1:n-k)$	87.506	96.876	99.213	99.803
$(1 - \frac{1}{2^{k+2}}) \times 100$	87.5	96.875	99.219	99.805

$n = 17$			
k	9	11	13
$S^{LPT}(1:n-k) = S^{RLPT}(1:n-k)$	99.950	99.988	99.997
$(1 - \frac{1}{2^{k+2}}) \times 100$	99.951	99.988	99.997

Table A.4: Relative number (in %) of instances so that
$S^{LPT}(1:n-k) = S^{RLPT}(1:n-k)$ for $n = 17$ and $k \in \{1, 3, 5, 7, 9, 11, 13\}$

$n = 51$				
k	1	3	5	7
$S^{LPT}(1:n-k) = S^{RLPT}(1:n-k)$	87.502	96.875	99.216	99.805

$n = 51$			
k	9	11	13
$S^{LPT}(1:n-k) = S^{RLPT}(1:n-k)$	99.953	99.988	99.997

Table A.5: Relative number (in %) of instances so that
$S^{LPT}(1:n-k) = S^{RLPT}(1:n-k)$ for $n = 51$ and $k \in \{1, 3, 5, 7, 9, 11, 13\}$

$n = 101$				
k	1	3	5	7
$S^{LPT}(1:n-k) = S^{RLPT}(1:n-k)$	87.491	96.872	99.220	99.806

$n = 101$			
k	9	11	13
$S^{LPT}(1:n-k) = S^{RLPT}(1:n-k)$	99.951	99.987	99.997

Table A.6: Relative number (in %) of instances so that
$S^{LPT}(1:n-k) = S^{RLPT}(1:n-k)$ for $n = 101$ and $k \in \{1, 3, 5, 7, 9, 11, 13\}$

Bibliography

[AL97] Aarts, E. H. L.; Lenstra, J. K. (Eds): Local Search in Combinatorial Optimization. John Wiley & Sons, New York (1997).

[ABN92] Arnold, B. C.; Balakrishnan, N.; Nagaraja, H. N.: A First Course in Order Statistics. John Wiley & Sons, New York (1992).

[Bła96] Błażewicz, J.: Scheduling Computer and Manufacturing Processes. Springer, Berlin (1996).

[BM08] Boettcher, S.; Mertens, S.: Analysis of the Karmarkar-Karp differencing algorithm. *The European Physical Journal B 65*, No. 1, 131-140 (2008).

[Bru95] Brucker, P.: Scheduling Algorithms. Springer, Berlin (1995).

[BCS74] Bruno, J.; Coffman, E. G., Jr.; Sethi, R.: Scheduling independent tasks to reduce mean finishing time. *Communications of the ACM 17*, No. 7, 382-387 (1974).

[Cof73] Coffman, E. G., Jr.: On a conjecture concerning the comparison of SPT and LPT scheduling. Technical Report #140, Computer Science Department, Pennsylvania State University, Pennsylvania (1973).

[CFL84] Coffman, E. G., Jr.; Frederickson, G. N.; Lueker, G. S.: A note on expected makespans for largest first sequences of independent tasks on two processors. *Mathematics of Operations Research 9*, No. 2, 260-266 (1984).

[CGJ78] Coffman, E. G., Jr.; Garey, M. R.; Johnson, D. S.: An application of bin-packing to multiprocessor scheduling. *SIAM Journal on Computing 7*, No. 1, 1-17 (1978).

[CG85] Coffman, E. G., Jr.; Gilbert, E. N.: On the expected relative performance of list scheduling. *Operations Research 33*, No. 3, 548-561 (1985).

[CL84] Coffman, E. G., Jr.; Langston, M. A.: A performance guarantee for the greedy set-partitioning algorithm. *Acta Informatica 21*, No. 4, 409-415 (1984).

[CL91] Coffman, E. G., Jr.; Lueker, G. S.: Probabilistic Analysis of Packing and Partitioning Algorithms. John Wiley & Sons, New York (1991).

[CS76] Coffman, E. G., Jr.; Sethi, R.: Algorithms minimizing mean flow time: schedule-length properties. *Acta Informatica 6*, 1 -14 (1976).

[CMM67] Conway, R. W.; Maxwell, W. L.; Miller, L. W.: Theory of Scheduling. Addison-Wesley Publishing Company, Reading (Massachusetts) (1967).

[CKW92] Csirik, J.; Kellerer, H.; Woeginger, G.: The exact LPT-bound for maximizing the minimum completion time. *Operations Research Letters 11*, Issue 5, 281-287 (1992).

[FD81] Friesen, D. K.; Deuermeyer, B. L.: Analysis of greedy solutions for a replacement part sequencing problem. *Mathematics of Operations Research 6*, No. 1, 74-87 (1981).

[DFL82] Deuermeyer, B. L.; Friesen, D. K.; Langston, M. A.: Scheduling to maximize the minimum processor finish time in a multiprocessor system. *SIAM Journal on Algebraic and Discrete Methods 3*, No. 2, 190-196 (1982).

[Doe09] Dörnfelder, M.: Penalty Methods in Discrete Optimization: On the Maximum Number of Threshold Parameters. Doctoral Thesis, Faculty of Mathematics and Informatics, Friedrich-Schiller-University Jena, Germany (2009).

[DSV97] Domschke, W.; Scholl, A.; Voß, S.: Produktionsplanung. Springer, Berlin (1997).

[FM87] Fischetti, M.; Martello, S.: Worst-case analysis of the differencing method for the partition problem. *Mathematical Programming 37*, No. 1, 117-120 (1987).

[FR86] Frenk, J. B. G.; Rinnooy Kan, A. H. G.: The rate of convergence to optimality of the LPT rule. *Discrete Applied Mathematics 14*, 187-197 (1986).

[FR87] Frenk, J. B. G.; Rinnooy Kan, A. H. G.: The asymptotic optimality of the LPT rule. *Mathematics of Operations Research 12*, No. 2, 241-254 (1987).

[Fri84] Friesen, D. K.: Tighter bounds for the Multifit processor scheduling algorithm. *SIAM Journal on Computing 13*, No. 1, 170-181 (1984).

[FL86] Friesen, D. K.; Langston, M. A.: Evaluation of a Multifit-based scheduling algorithm. *Journal of Algorithms 7*, No. 1, 35-59 (1986).

[GJ79] Garey, M. R.; Johnson, D. S.: Computers and Intractability: A Guide to the Theory of NP-Completeness. W. H. Freeman, San Francisco (1979).

[Gra66] Graham, R. L.: Bounds for certain multiprocessing anomalies. *Bell Systems Technical Journal 45*, No. 9, 1563-1581 (1966).

[Gra69] Graham, R. L.: Bounds on multiprocessing timing anomalies. *SIAM Journal on Applied Mathematics 17*, No. 2, 416-429 (1969).

[GLL++79] Graham, R. L.; Lawler, E. L.; Lenstra, J. K.; Rinnooy Kan, A. H. G.: Optimization and approximation in deterministic sequencing and scheduling: a survey. *Annals of Discrete Mathematics 5*, 287-326 (1979).

[Hal97] Hall, L. A.: Approximation algorithms for scheduling. In: Hochbaum, D. S. (Ed.): Approximation Algorithms for NP-Hard Problems. PWS Publishing Company, Boston (1997).

[Hau89] Haupt, R.: A survey of priority rule-based scheduling. *OR Spektrum 11*, 3-16 (1989).

[KK82] Karmarkar, N.; Karp, R. M.: The differencing method of set partitioning. Technical Report UCB/CSD 82/113, Computer Science Division, University of California, Berkeley (1982).

[Kar72] Karp, R. M.: Complexity of computer computations. In Miller, R. E. ; Thatcher, J. W. (Eds.): Reducibility Among Combinatorial Problems, 85-103, Plenum Press, New York (1972).

[KT05] Kleinberg, J.; Tardos, E.: Algorithm Design. Addison-Wesley, Boston (2005).

[LM88] Lee, C. -Y.; Massey, J. D.: Multiprocessor scheduling: combining LPT and Multifit. *Discrete Applied Mathematics 20*, Issue 3, 233-242 (1988).

[Lue87] Lueker, G. S.: A note on the average-case behavior of a simple differencing method for partitioning. *Operations Research Letters 6*, Issue 6, 285-287 (1987).

[Mer99] Mertens, S.: A complete anytime algorithm for balanced number partitioning. 1999. <http://arXiv.org/abs/cs.DS/9903011>.

[Mic04] Michiels, W. P. A. J.: Performance Ratios for the Differencing Method. Ph. D. Thesis, Faculty of Mathematics and Computer Science, TU Eindhoven, The Netherlands (2004). <http://alexandria.tue.nl/extra2/200410775.pdf>.

[MR05] Moreira, N.; Reis, R.: On the density of languages representing finite set partitions. *Journal of Integer Sequences 8*, Article 05.2.8 (2005).

[Rin86] Rinnooy Kan, A. H. G.: An introduction to the analysis of approximation algorithms. *Discrete Applied Mathematics 14*, 171-185 (1986).

[Ste02] Steger, A.: Diskrete Strukturen 1. Springer, Berlin (2002).

[Tsa92] Tsai, L. -H.: Asymptotic analysis of an algorithm for balanced parallel processor scheduling. *SIAM Journal on Computing 21*, No. 1, 59-64 (1992).

[Tsa95] Tsai, L. -H.: The modified differencing method for the set partitioning problem with cardinality constraints. *Discrete Applied Mathematics 63*, No. 2, 175-180 (1995).

[Wal01] Walter, W.: Analysis 1. Springer, Berlin (2001).

[Woe97] Woeginger, G. J.: A polynomial-time approximation scheme for maximizing the minimum machine completion time. *Operations Research Letters 20*, Issue 4, 149-154 (1997).

[Yak96] Yakir, B.: The differencing algorithm LDM for partitioning: a proof of a conjecture of Karmarkar and Karp. *Mathematics of Operations Research 21*, No. 1, 85-99 (1996).

[Yue90] Yue, M.: On the exact upper bound for the Multifit processor scheduling algorithm. *Annals of Operations Research 24*, 233-259 (1990).